AF808672

WILLIAM STANLEY JEVONS

LOGIC

Elibron Classics
www.elibron.com

Elibron Classics series.

ISBN 1-4021-5987-0 (paperback)
ISBN 1-4021-0422-7 (hardcover)

This Elibron Classics Replica Edition is an unabridged facsimile of the edition published in 1894 by Macmillan and Co., London and New York.

SCIENCE PRIMERS, *edited by*

Professors Huxley, Roscoe, *and*

Balfour Stewart.

LOGIC.

RICHARD CLAY AND SONS, LIMITED,
LONDON AND BUNGAY.

First Edition, May 1876.
Reprinted, December 1876, 1878, 1880, 1884, 1886, 1889, 1891, 1894.

Science Primers

LOGIC

BY

W. STANLEY JEVONS, M.A., LL.D., F.R.

WITH ILLUSTRATIONS

London
MACMILLAN AND CO.
AND NEW YORK
1894

CONTENTS.

SCIENCE PRIMERS.

LOGIC.

I.—INTRODUCTION.

1. MONSIEUR JOURDAIN, an amusing person in one of Molière's plays, expressed much surprise on learning that he had been talking prose for more than forty years without knowing it. Ninety-nine people out of a hundred might be equally surprised on hearing that they had long been converting propositions, syllogizing, falling into paralogisms, framing hypotheses and making classifications with genera and species.

If asked whether they were logicians, they would probably answer, No! They would be partly right; for I believe that a large number even of educated persons have no clear idea what logic is. Yet, in a certain way, every one must have been a logician since he began to speak.

It may be asked:—If we cannot help being logicians, why do we need logic books at all? The answer is that there are logicians and logicians. All people are logicians in some manner or degree; but unfortunately many people are bad ones, and suffer harm in consequence. It is just the same in other matters. Even if we do not know the meaning of the name, we are all **athletes** in some manner or degree. No one can climb a tree or get over a gate without being more or less an athlete. Nevertheless, he who wishes to do these actions really well, to have a strong muscular frame, and thereby to secure good health and personal safety, as far as possible, should learn athletic exercises under a skilful teacher.

2. To be a good logician is, however, far more valuable than to be a good athlete; because **logic teaches us to reason well, and reasoning gives us knowledge, and knowledge, as Lord Bacon said, is power.** As athletes men cannot for a moment compare with horses or tigers or monkeys. Yet, with the power of knowledge, men tame horses and shoot tigers and despise monkeys. The weakest framework with the most logical mind will conquer in the end, because it is able to foresee the future, to calculate the results of actions, to avoid mistakes which might be fatal, and to discover the means of doing things which seemed impossible. If such little creatures as ants had better brains than men, they would either destroy men or make them into slaves.

3. It is true that we cannot use our eyes or ears without getting some kind of knowledge, and the brute animals can do the same. But **what gives power is the deeper knowledge called Science.** People may see, and hear, and feel all their lives without really learning the nature of things they see. But reason is the mind's eye, and enables us to see why things are, and when and how events may be made to happen or not to happen. The logician endeavours to learn exactly what this reason is which makes the power of men. We all, as I have said, must reason well or ill, but logic is the science of reasoning and enables us to distinguish between the good reasoning which leads to truth, and the bad

reasoning which every day betrays people into error and misfortune.

II.—HOW WE COMMONLY REASON.

4. The common way in which we reason is to expect that things will happen as they have happened before in like circumstances. Seeing a bright flash of lightning, I expect thunder to follow, because it has followed bright flashes of lightning in previous cases. When a bright yellow round fruit is offered to me I believe it to be an orange and eat it without hesitation, because fruit of exactly the same appearance had been eaten before without harm. The gold of Australia was discovered by this simple mode of reasoning. A man named Hargreaves remarked that the mountains of New South Wales were like those of California, where he had been digging gold, and he reasoned that being like in some respects, they ought to be like in other respects, and should contain gold. On making some trials he found that he was correct.

5. But in this simple way of reasoning from like to like we may often deceive ourselves. When the things which we believe to be like each other are really so, no harm is done; but things which seem to be like may be different: two kinds of fungus or two kinds of fruit may so closely resemble each other that I may not notice the difference; yet one kind may prove to be wholesome to eat and the other poisonous. It is even possible that what looks exactly like an orange might be some new sort of fruit, and not an orange at all.

People are so accustomed to use blankets to make themselves warm that they are surprised to see blankets used to keep ice cold, and to prevent it from melting. Expecting that the same thing will have the same effect, they think that a blanket must make ice warm. But this would not really be a similar effect. What a blanket always does is to prevent heat passing from one side to the other. Thus it keeps the heat of the body from passing into the colder air around, and it keeps the heat of the air from passing into the colder ice. Housemaids, in trying to make a fire burn, sometimes reason badly. They stick the poker among the coals and leave it there, seeming to have a belief that the mere presence of the poker helps the fire to burn, because on some previous occasions the fire had burnt better when the poker was in it. They do not observe that the poker is only useful when so placed as to raise the coals and allow the air to enter freely.

6. The truth is that only when things really are alike can we expect them to behave alike. The same causes have the same effects; but the difficulty is to know when the causes are the same. To ascertain this requires more careful reasoning than we commonly use. We need to discover what things go with other things always and everywhere as far as we can observe. We have to find out what are called **the general laws showing what things will happen under given circumstances.** A fire sometimes burns and sometimes does not burn. Then the circumstances must be different; for a fire has no will; and if one fire be laid and lighted exactly like another, it ought to burn like it. We must find out what things always favour the burning, such as the presence of abundant air, and the absence of moisture, and of any body which can carry much heat away. We shall thus learn that a cold poker put into a fire in one way will do more harm than good, by carrying away heat; but, put in differently, it will do more good than harm, by admitting air and quickening combustion.

7. **A general law of nature is something which is true of many objects, and science is made up of such laws.** After reflecting a little, we shall see that logic ought to teach us to do two different things with respect to the laws of nature, namely, how to discover them, and how to use them when discovered. By **inductive reasoning**, as it is called, we ascertain what is true of many different things. Our eyes, and ears, and other senses tell us what happens around us, and then by proper reasoning we may often discover the laws of nature, in consequence of which they happen. Observing that clouds, rain, snow, hail, dew, mist, and fogs, consist of water, which seems to come out of the air, we may, by a proper course of inquiry, discover that all moist air, when cooled in a certain degree, produces particles of water. We find that there is something the same in the causes of all these things.

8. By **deductive reasoning** we do just the opposite, and from any law of nature we infer what will happen in consequence of it. **To infer is to find out what will be true if something else is true.** Knowing that moist air when cooled produces particles of water, I may infer that an iced bottle of wine will in summer become covered with dew. Philosophers have discovered by induction that all bodies tend to fall towards the earth like stones; then by deduction I can infer that the moon must tend to fall towards the earth. It may seem as if all the difficulty of reasoning lies in discovering laws by induction, and that we must certainly learn to discover the laws before we learn how to use them. The fact, however, is that we cannot possibly understand inductive reasoning unless we previously understand deductive reasoning.

9. **Before we can be said to know properly what a law of nature means, we must be able to see what it**

leads to, that is, to infer its consequences. I cannot tell whether a law is true or not unless I see whether it agrees with what happens in nature. When philosophers came to the conclusion that all material bodies tend to fall towards the earth, they ought to have been able to foresee that the moon, being a material body, would tend to fall towards the earth, so as to inquire whether this was true or not. I shall afterwards show more fully that it is really by the use of deductive reasoning that we perform inductive reasoning. We will now proceed at once to consider what deductive reasoning consists in.

III.—WHAT IS DEDUCTIVE REASONING?

10. Let us take a simple case of reasoning, an argument, as it is often called, and consider in what way it is constructed. When we see a particular kind of white and pink fungus, and pluck it, because we believe it to be a mushroom, and we know that all mushrooms are good to eat, we certainly reason by an argument, which may be thus fully stated:—

All mushrooms are good to eat;
This fungus is a mushroom;

Therefore, this fungus is good to eat.

Here are three sentences which state three different facts; but when we know the two first facts, we learn or gather the third fact from the other two. When we thus learn one fact from other facts, we infer or reason, and we do this in the mind. Reasoning thus enables us to ascertain the nature of a thing without actual trial. If we always needed to taste a thing before we could know whether it was good to eat or not, cases of poisoning would be alarmingly frequent. But the appearance and peculiarities of a mushroom may be safely learned by the eye or the nose, and reasoning upon this information and the fact already well-known, that mushrooms are good to eat, we arrive without any danger or trouble at the conclusion that the particular fungus before us is good to eat. **To reason, then, is to get some knowledge from other knowledge.**

11. Let us now examine more carefully the parts of which this argument about mushrooms is made up. There are three sentences, which put facts before us, and are therefore called **Propositions**. The first proposition tells us that "All mushrooms are good to eat," or, what means exactly the same, "All mushrooms are things good to eat." This proposition has three principal parts. There are two descriptions of things compared together, namely, "mushrooms" and "things good to eat." These kinds of things are mentioned, of course, by their names, and, as the name mushroom is at one end of the proposition, and things good to eat at the other end, these names are called the **terms**, or ends of the proposition. They are connected together by the little verb "are," which is called the **copula** or link. There remains, indeed, the little adjective "all," which tells us how many of the mushrooms are good to eat; in the case of other things it might be few, or many, or none. In this case it is all, and we may call this the **sign of quantity.**

The other propositions are made up nearly in the same way. Thus, in "This fungus is a mushroom," we observe two terms, namely, "this fungus," and "mushroom," which are connected together by the copula "is." In the third proposition, which we drew from the other two, the terms, "this fungus" and "thing good to eat," are found over again, with the copula "is." It will be observed that each term is used twice over in the argument; "this fungus" occurs in the second and third propositions; "mushroom" in the first and second; and "thing good to eat" in the first and third. We learn from our examination, that an argument of this kind consists of three propositions and of three terms, and that each proposition is made by joining two terms. **When we join terms together we make a proposition; when we join propositions together we make an argument, or piece of reasoning.**

12. We should generally get nothing but nonsense if we were to put together any terms and any propositions, and to suppose that we were reasoning. To produce a good argument we must be careful to obey certain rules, which it is the purpose of Logic to make known. But, in order to understand the matter perfectly, we ought first to learn exactly what a term is, and how many kinds of terms there may be; we have next to learn the nature of a proposition, and the different kinds of propositions. Afterwards we shall learn how one proposition may by reasoning be drawn from other propositions in the kind of argument called the syllogism. Thus, **there are three parts of Deductive Logic, which treat of Terms, Propositions, and Syllogisms.** Terms and propositions are, indeed, merely the tools which we use in reasoning: but we cannot learn a trade unless we begin with learning the use of the tools employed in it. Hence we shall begin by considering the different kinds of terms and propositions before we go on to the syllogism.

IV.—THE DIFFERENT KINDS OF TERMS OR NAMES.

13. As we have already learnt, terms are the **names** of the things which we compare together in a proposition. Now names consist of what the grammar books call nouns, and a single term may con-

tain any number of nouns, substantive or adjective. Sometimes there is in each term only a single noun. Thus in "Diamonds are combustible," the first term is the single substantive "diamonds"; the second term is the single adjective "combustible." But a term will often be made of two or more nouns joined together in some way. The proposition "The Queen of England is the Empress of India" contains only two terms, but each of them is composed of two nouns, "Queen of England" being the first term, "Empress of India," the second. "The library of the British Museum is the greatest collection of books in the world." Here is a proposition containing fifteen words, yet it has only two terms; the first is "The library of the British Museum," in which we see two substantives, one adjective, two definite articles and one preposition; the second is "the greatest collection of books in the world," in which we see three substantives, one adjective, two articles, and two prepositions. A logical term, then, may consist of any number of nouns, substantive or adjective, with the articles, prepositions and conjunctions required to join them together; still it is only one term if it points out, or makes us think of a single object, or collection, or class of objects. But there are several different kinds of terms, which we must next consider.

14. Sometimes a term points out only a single person or thing, as "The Queen of England," "The British Museum," "Pompey's Pillar." By the Queen of England we mean the present reigning Queen Victoria, and there is, of course, only one Queen Victoria. There is only one British Museum, and one single great obelisk called Pompey's Pillar. Hence terms of this kind are called **singular terms**, because each term is the name only of a single thing.

15. The greater number of terms used in writing and speaking are, however, not singular but **general**

terms. They are the names of things of which many exist. Thus "shilling" is not the name of any one single thing, like "Pompey's Pillar." There are many millions of things, each of which can be called a shilling; and when I say that "all shillings are made of a mixture of silver and copper," I mean to state this of any and all the shillings. In the same way "horse" is the name of any one of millions of horses which may exist in the world. The number of things denoted by a general term may vary from two or three to numbers exceeding anything that we can conceive. "Present King of Siam" is the general term for either of the two existing kings of that country; "House of Parliament" is the general name either of the Lords or the Commons. "Grain of sand," is the name of any one of many billions, or even trillions of little particles; and "particle of matter" is a still wider general name; for all substances which exist in the universe are composed of minute particles of matter.

16. It may be remarked, indeed, that, as even a single thing, like Pompey's Pillar, is made up of many portions of matter, the name of the whole must be the name of all the parts. The continent of Asia is made up of many plains, lakes, mountains and rivers. Polynesia is the name of an immense number of islands scattered about the Pacific Ocean. Nevertheless, each of these things is a single whole. There are not two Pompey's Pillars in existence, nor two Asias, nor two Polynesias. Hence each of these terms is a singular term, not a general term, and a singular term may be the name of many things, provided they are all put together into one single group or collection. Polynesia is the name not of any one island, but of a great many islands in the Pacific Ocean. Such a term is called a **Collective term**, because it is the name of many things collected into one whole.

Library is the collective name for many books put together; constellation, of many stars; crowd, of many people.

17. I have said that a general name is the name of many things; but then it is the name of any one of those things separately from the others. Thus "island" is the name of any one of the thousands of small pieces of land making up Polynesia. Island, then, is a general term; Polynesia is a collective and singular term. The British Museum Library is the name of a great collection of books, not of any one of those books; it is, therefore, a collective term, and is also singular. There are, however, a great many collections of books of various size in the world, so that the term "Library," though it is collective as regards the books in any particular library, is yet general, because it is the name of any such collection. We thus see that the same term may be at once collective and singular, or collective and general; but **we must always take great care to avoid confusing collective terms with general terms.**

18. There is another difference between terms which it is not so easy to understand. Many terms are the names of solid objects, which we can touch or move about, and which exist by themselves, like a half-crown, a writing slate, or a brick house. Such terms are called **Concrete terms**, and they include most names which may be put in the plural; thus we may speak of half-crowns, brick houses, mountains, planets, particles of matter, and so on. All these are concrete terms.

Abstract terms, on the contrary, are names, not exactly of things, but of qualities which belong to things, as the thickness of a half-crown, the colour of a slate, the magnitude of a house, the elevation of a mountain. We cannot separate the thickness of a half-crown from the half-crown, as we can separate one half-crown from another. Every object has many qualities; a half-crown, besides thickness, has weight, solidity, colour, ductility, malleability, fusibility, conductibility, and many other qualities, so that each of these terms is an abstract one. Properly speaking, an abstract term cannot be put into the plural. We ought not to speak of solidities, ductilities, fusibilities, these being perfectly abstract terms. It is true that we often speak of colours, weights, magnitudes; but it is probable that we then make the terms concrete. Altogether, there is much confusion between abstract and concrete terms, and the difference between them is not well understood. It will be sufficient to remember that **a concrete term is the name of a thing; an abstract term is the name of a quality of a thing.**

19. We must now ascertain the difference between **positive and negative terms.** As a general rule we give a name to a thing because it has a certain quality. We call a house "a brick house," because it is made of bricks; black-lead is so called because it is black, and looks like lead. But in other cases we give a name to a thing for the opposite reason, because it has not got a certain quality. Thus we call a feat impossible, because it cannot be done; a speech is unparliamentary when it does not agree with the rules of parliamentary debate; an immense distance means a distance which has not been measured; an uneven surface is one not possessing evenness; unfiltered water is water which has not been filtered. All these are negative terms. We may usually know a negative term by its beginning with one of the little syllables un-, in-, a-, an-, non-, or by its ending with -less. Thus unfavourable, indivisible, amorphous, anonymous, non-metallic, useless, are negative terms. But there are also many terms which

may be said to serve as negative terms, although they have no such mark at the beginning or end. When a piece of metal can be hammered out into a thin plate we call it malleable; when it cannot be so hammered out, it might be called immalleable; but this word has seldom been used, and we generally call such a piece of metal brittle. Thus "brittle" serves as the negative term of "malleable." Similarly, opaque is the negative of transparent, false of true, dry of moist, rough of smooth, and so on. When we are speaking of written or spoken compositions, verse is the negative of prose, and prose the negative of verse, unless, indeed, Monsieur Jourdain was right in thinking that he could get a love-letter written neither in verse nor in prose.

20. If the English language were a perfect one, every term ought to have a negative term exactly corresponding to it, so that all adjectives and nouns would be in pairs. Just as convenient has its negative inconvenient; metallic, non-metallic; logical, illogical; and so on; so blue should have its negative non-blue; literary, non-literary; paper, non-paper. But many of these negative terms would be seldom or never used, and, if we happen to want them, we can make them for the occasion by putting not-, or non-, before the positive term. Accordingly, we find in the dictionary only those negative terms which are much employed. When we are speaking in England of those belonging to Christian sects, a Churchman means one belonging to the Church of England. Those who, being Christians, are not Churchmen, are called Dissenters, so that the term Dissenter serves as the negative of Churchman. But we have no separate names for those who are not-Wesleyans, or not-Methodists, or not-Baptists.

Sometimes the same word may seem to have two or even more distinct negatives. There is much difference between undressed and not-dressed, that is "not in evening dress." Both seem to be negatives of "dressed," but this is because the word has two distinct meanings.

21. Mistakes frequently arise from not observing the distinction between negative terms which indicate the complete absence of some quality, and comparative or opposite terms which only mean various degrees of the property. Thus the term "small" is not really the negative of "large," because there may be things which are neither large nor small, that is, are of medium size. The negative of large is not-large, which includes both medium and small; similarly, the negative of small is not-small, which includes both medium and large. So with warm and cold, light and dark, heavy and light; these are not pairs of positive and negative terms, unless by cold we mean the entire absence of warmth, by dark the entire absence of light, and so on, as we seldom do. We never can make anything so cold that it contains no heat at all; the question is altogether one of degree. Thus the word "hot," as we generally use it, does not mean "possessing heat," the negative of which would be "not possessing heat," but "possessing more than medium heat," the negative of which is "not possessing more than medium heat," and includes both things which are of medium temperature and those which would be called cold. If then a person denies that a thing is hot, he ought not to be understood as asserting that it is cold, for it may be just short of being hot, and yet not cold.

V.—THE FULL MEANING OF TERMS.

22. We cannot really form a clear notion of what a concrete term means unless we observe that there are two different kinds of meanings, namely, **the things to which the term is applied, and the**

qualities of those things in consequence of which it is applied. When I see a large peculiar-shaped iron structure floating on the water with masts and sails, I call it a ship, because it is evidently adapted to sailing and conveying goods and passengers. Every other structure having the same general appearance and purpose I also call ship, and if I were asked Why, I should have to answer as best I could, that every large vessel made to move through the water easily and convey things is a ship. Whenever then, I call a thing a ship I must mean that it has these peculiarities; it is these circumstances which make it a ship, and lead me to use the name ship; so that the word means that the thing to which it is applied is made to move easily through the water, and so forth. But, on the other hand, the name ship is the name of the thing, and there are a great many particular ships, such as the *Great Britain*, the *Great Eastern*, the *Challenger*, the *Castalia*, the *Minotaur*, the *Vanguard*.

In reality every ordinary general term has a double meaning: it means the things to which it is **applied**, for instance, the particular ships named: it also means in a totally different way, the qualities and peculiarities **implied** as being in the things. Logicians say that the number of things to which a term applies is the **extension** of the term; while the number of qualities or peculiarities implied is the **intension**.

23. When we compare together terms which are partly different and partly the same, we shall find that they have various degrees of extension and intension. Take, for instance, the term "ship" and compare it with "steam-ship." There are evidently many more ships than there are steam-ships, because we have in the meaning of the latter term to exclude sailing ships. Hence in putting steam before ship we have greatly reduced the extension of the term. But

we have increased its intension, because steam-ship means all that ship does, and more, for it means that the ship is moved by steam power. Put another word before it and compare screw-steam-ship with steam-ship, and we find we have again reduced the extension, by putting out of sight the steam-ships yet propelled by paddles; these are comparatively few in the present day, so that we have not made any very great difference; but we have nevertheless increased the intension of meaning considerably, because we know precisely in what way a thing called a screw-steam-ship is moved. War-screw-steam-ship is a still narrower term, that is, has much less extension, because it now applies only to those ships owned by a government for war purposes; but this makes an addition to the intension or the circumstances and qualities implied. British-war-screw-steam-ship is in like manner a still narrower term, and we might go on further specifying that it is iron clad, that it is in commission, is in the Channel Fleet. We have thus narrowed the extension so much that there may only be half-a-dozen ships to which our description will now apply. If we add that it bears the Admiral's flag, this removes all but a single ship, so that the extension is reduced to the least possible. At the same time the intension becomes very great, and if we happen to be acquainted with the ship, and to have heard much about it, all the knowledge which we have of the ship is suggested by the name.

VI.—THE CORRECT USE OF WORDS.

24. In endeavouring to reason correctly, there is nothing more necessary than to use words with care. **The meaning of a word is that thing which we think about when we use the word,** and

which we intend other people to think about when they hear it pronounced, or see it written. We can hardly think at all without the proper words coming into the mind, and we can certainly not make known to other people our thoughts and arguments unless we use words. Yet there is no more common source of mistakes and bad reasoning than the confusion which arises between the different meanings of the same word.

25. Take for instance the word "church." It may, no doubt, be said to mean the solid building of stone or brick, to which people go to worship, and when used in this sense there can seldom be any important mistake. But it is also common to speak of the Church as meaning the whole body of people who worship in a particular manner, and have the same creed and ritual. Thus there is the Church of England, the Church of Rome, the Greek Church, the Free Church of Scotland, and so forth. When we say a person has gone over to the Church of Rome, we do not mean that he has gone bodily to Rome, but that he has simply changed his belief. Each different sect too speaks of the Church as meaning their own Church, so that two people arguing together and speaking of the Church may mean totally different Churches.

26. There is, however, a still more serious confusion in the meanings of the word, because the bishops, clergy, and other authorities of the Church, being the most prominent members of it, and governing, representing, and expressing the opinions of the Church, often come to be spoken of as if they were the Church. Properly speaking, the congregation who attend worship have as much right to be considered part of the Church as the clergy, and they have, to a certain extent, the right of electing officers, of deciding questions about the building, and so forth.

But, if we include the congregation, how are we to decide who shall be counted as proper members? Not everybody who goes inside a church door can be called a member of the Church. For some purposes we should include only regular communicants; for other purposes those who have been baptized, and confirmed, and not excommunicated; many overlook the confirmation, and there may be persons who, without having been even baptized, consider themselves members of a Church, because they have regularly attended worship, and have subscribed towards the expenses. Even while we are discussing the word "church," it is difficult to avoid speaking more in reference to some one Church, for instance, the Church of England, than to other Churches. We must remember that the Wesleyan, Baptist, Roman Catholic, and many other Churches, take much care to ascertain and settle, who do, and who do not belong to their particular Churches.

27. In many cases the meanings of a word are so distinct that they cannot really lead us into more than a momentary misapprehension, or give rise to a pun. A rake may be either a garden implement, or a fast young man; a sole may be a fish, or the sole of the foot; a bore is either a tedious person, a hole in a cannon, or the sudden high wave which runs up some rivers when the tide begins to rise; diet is the name of what we eat daily, or of the Parliament which formerly met in Germany and Poland; ball is a round object, or a dance. In some cases a word is really a different word in each of two or three meanings, and comes from quite different words in other languages. Thus, *bale* is the name of a bundle or package of goods, and seems to be derived from the same French or Latin words as *ball;* but *bale* is also an old name for evil, calamity, or sorrow, and in this meaning comes from an altogether distinct root. The

corn which we eat is the Latin *granum*, but a corn or horn on the foot is the Latin *cornu*. Bill means either William, a document, or a hooked object, for instance, the bill of a bird. In each case it is really a different word similarly spelt. From such confusions of words puns and humorous mistakes may arise, but hardly any important errors.

28. In most cases a word changes its meaning by degrees, and we use it for anything which is close to, or connected with, the first meaning. A bench means a board to sit on, but "the bench" is a common expression for the row of magistrates sitting on the bench. A board means a broad flat piece of wood, but being often used to support the dishes at a meal, people speak of the food itself as the board. Again, because a small meeting of men often sit round a table for convenience, they are sometimes called a board, as in the Board of Trade, and a small meeting-room is very commonly called a board-room.

29. Any word which has two or more meanings, and is used in such a way that we are likely to confuse one meaning with another, is said to be **ambiguous**, or to have the quality of **ambiguity**. By far the greater number of words are ambiguous, and it is not easy to find many words which are quite free from ambiguity. Whether we are writing, or reading, or speaking, or merely thinking, we should always be trying to avoid confusion in the use of words, but no one can hope to avoid making blunders and falling into occasional fallacies, as we shall learn in a later part of this Primer.

30. In many important cases it seems almost impossible to decide exactly what a name means. House, for instance, has a great many meanings. No doubt, it first meant any kind of roofed building in which people live; but, as the shelters made for cows much resembled houses, they were called cow-houses;

and we speak now of ice-houses, tool-houses, green-houses, hot-houses, bathing-houses, wash-houses, and many other kinds of houses, in which no living beings remain long. Counting-houses, again, being the chambers in which men conduct business, we often speak of the house instead of the men, just as we speak of the bench instead of the magistrates. Thus a commercial house means a firm, or partnership, or company doing business together. As members of Parliament need chambers to meet and debate in, there is the House of Lords, and the House of Commons, and it is usual to speak of "The House" meaning the collection either of Lords or Commons who happen to be present. Here again there is ambiguity; for the House of Commons may mean either the members who happen to be in the House at any particular moment, or the whole body of 652 members whose right and duty it is to be there when the Speaker is sitting.

31. Even beyond all these varieties of meaning, there is further uncertainty as to what house means when it is merely a dwelling-house. Houses are of various sizes, and if a family live in a building having only one single chamber, it would be a house. Legally speaking, the head of the family would be a householder. If several poor families divide a large house between them, each taking one or two rooms, we still speak of the whole building as one house. Nevertheless, it is for practical purposes made into several houses. If a single room standing alone makes a house, as in the case of many cottages, why should not single rooms inhabited by different families under the same roof make different houses? Whether there is or is not an outer front door is not a matter of real importance. If in this way we follow out our use of the word house, we shall find that we cannot give a satisfactory account of it.

VII.—HOW AND WHY WE CLASSIFY THINGS.

32. The larger number of terms, as we have learnt in Art. 15, are the names, not of single objects, but of many objects, or rather of any one of many objects. "Man" is the name of any one of many hundreds of millions of men, living or dead. We have hitherto called such names, **general names** or terms; but we may now say that they are the **names of classes of things**, provided that we take great care to ascertain exactly what we mean by a class.

We class things together whenever we observe that they are like each other in any respect, and therefore think of them together. Milk, chalk, snow, meerschaum, paper, mist, spray, foam of the sea, pearls, and white lead, are very different things in most points, but they all agree in being white. Together with many other substances and things, they are put together in thought into **the class of white things.** In this case the resemblance is only in regard to colour; but in other cases there may be many points of resemblance.

The class of things called "pens," for instance, includes things made of quills, reeds, steel, gold, silver, glass, and some other substances; the forms of pens also differ; nevertheless, they are all like each other in being made to hold fluid ink, and to spread it over paper.

33. There is nothing more useful than to be able to classify things correctly and easily, and to form exact general notions about them. **So far as things are exactly alike, whatever is true of one thing will be true of the others, which so resemble each other. When we classify things correctly, we ascertain the exact**

nature and degree of their resemblances, and record the information we have gained in the briefest and most convenient form. Our knowledge is increased to the utmost, and, instead of being obliged to remember an immense number of disconnected facts, we have only to comprehend a comparatively small number of general truths. To take a very simple case, we class together white things because they all act in the same way with respect to light. Linen, snow, chalk, cloud, and porcelain, are exceedingly different in other respects, so that it is only with regard to light that we expect the same fact to be true of all of them. Those who walk over a large extent of snow when the sun is shining, find their eyes painfully affected by the glare of the light reflected from the snow. They might expect therefore that the same effect would follow from walking over a large extent of ground covered with white chalk, white dust, or white linen laid out to bleach in the sun. Again, when we want to reflect light we shall know that we ought to use white substances; in a dark room we should have a white ceiling, and white paper or paint on the walls. If there be walls in front of a window, they should be built of white-faced bricks, or covered with white-wash, if we want additional light in the room, and white bricks are often used for this purpose at the present day. Sometimes, too, whiteness will assist us in avoiding the effects of the excessive intensity of the sun's rays; people in tropical countries wear white clothes and hats with this object, and roofs are sometimes white-washed in order that they may absorb less of the sun's heat. All these results follow from the one general truth or law that white things reflect rays of light.

34. The studies of botanists and other naturalists are chiefly directed to classifying plants and animals in the most perfect manner, because it is only by

classification that we can possibly remember or understand the characters of the immense numbers of living things. All kinds of grasses, including wheat, barley, oats, and other kinds of corn, belong to one very well-marked class. Any one having a moderate knowledge of botany can tell with ease whether a given plant is a kind of grass or not. Now the food both of men and brutes is chiefly derived from some sort of grass, and it is believed with much reason that no plant belonging to the class is poisonous. Hence a traveller in want of food in an uninhabited country might always eat the seeds of any kind of grass without fear. On the other hand, plants belonging to the order *Lobeliaceæ* should never be eaten, as most if not all of them are dangerously poisonous. The same may be said of the flowers and berries of plants belonging to the order of *Solanaceæ*, among which is the Deadly Night-shade. A good botanist would know, almost at a glance, that these and many other classes of plants were to be avoided, or very carefully used.

35. It is somewhat the same with classes of substances or living beings. The properties of the class "man" are exceedingly numerous. The surgeon who has well studied anatomy knows almost exactly the form and place of every bone, tendon, muscle, nerve, gland, or other organ. There are various circumstances in which one man may differ from another; these are in logic called **accidents**. An organ or muscle may be smaller or larger in one man than another; but it will be present in all, so that the possession of the organ is a **property** of the man. Chemical substances, again, have innumerable well-marked properties. If a chemist meets with a transparent colourless crystal, and decides by certain tests that it is composed of carbonate of lime, he knows at once how it will behave, if treated with

various acids, or if burnt in the fire; for he knows the properties which belong to all portions of carbonate of lime.

36. In classifying things, however, we must take great care not to be misled by outward resemblances. Things may seem to be very like each other which are not so. Whales, porpoises, seals, and several other animals live in the sea exactly like fish; they have a similar shape, and are usually classed among fish. People are said to go whale-fishing. Yet these animals are not really fish at all, but are much more like dogs and horses and other quadrupeds than they are like fish. They cannot live entirely under water and breathe the air contained in the water like fish, but they have to come up to the surface at intervals to take breath.. Similarly, we must not class bats with birds because they fly about; although they have what would be called wings, these wings are not like those of birds, and in truth bats are much more like rats and mice than they are like birds. Botanists used at one time to classify plants according to their size, as trees, shrubs, or herbs, but we now know that a great tree is often more really similar in its character to a tiny herb than it is to other great trees. A daisy has little resemblance to a great Scotch thistle; yet the botanist regards them as very similar. The lofty growing bamboo is a kind of grass, and the sugar-cane also belongs to the same class with wheat and oats.

37. In classifying a collection of objects, we do not merely put together into groups those which resemble each other, but we also often divide each larger class into smaller ones, in which the resemblance is more complete. Thus, the class of white substances may be divided into those which are solid and those which are fluid, so that we get the two minor classes of solid white, and fluid white substances. It is

desirable to have names by which to show that one class is contained in another, and accordingly we call the class which is divided into two or more smaller ones, the **genus**, and the smaller ones into which it is divided, the **species**. Solid white substance is a species of the genus white substance. If house be taken as a genus, then dwelling-house would be a species. But, when we like, we can again turn the species into a genus, by dividing it up a second time; thus, brick dwelling-house would be a species of the genus dwelling-house. This we might do again and again, getting, for instance, the still smaller species, new brick dwelling-house, large new brick dwelling-house, Elizabethan large new brick dwelling-house, and so on, almost without limit.

38. It is often a difficult question to decide how, in any particular case, we can best divide up a large class into smaller ones. The common way is to make as many species all at one step, as there are kinds of things belonging to the class, which we can think of at the time. Thus, we might divide boats into sailing-boats, steam-boats and row-boats. Beasts of burden might be divided into horses, mules, donkeys, camels, and elephants. Books might be divided into those which treat of History, Geography, Biography, General Literature, the Physical and Moral Sciences, the Arts, Political Economy, Theology, Poetry, Fiction, Periodical Publications, &c. But, in making such classifications, we are almost sure to fall into logical blunders.

39. In the first place the species or small classes are likely to overlap each other, unless we make the divisions with much care. If we divide the people of England into men, women, children, paupers, vagrants, blind, deaf and dumb, and foreigners, we commit several very evident blunders, because paupers, blind, deaf and dumb, as well as foreigners, must be either

men, women, or children, so that if they were counted once in that respect they ought not to be counted again as paupers, blind persons, &c. Vagrants are a kind of paupers, and often difficult to distinguish from them. Moreover, vagrants and foreigners may happen to be blind, or deaf and dumb. In dividing books, again, it will be found impossible to make any classification in which a book shall always belong to one species and only to one. The species will be sure to overlap. There may be books on the history of science which might be equally well placed in the class of histories, or in that of books on physical science. There may be books which are half biography, half history. Miss Martineau's "Tales on Political Economy," might be placed both in the class of fiction and in that of political economy. Nobody can be sure in which class any particular book will be found, and accordingly such classifications are not only logically bad ones, but they are of little use. Yet we find them employed in the catalogues of many libraries.

40. A second difficulty is, that, in planning such classifications, we can seldom be sure of making enough species to include all the things belonging to the genus. There may be beasts of burden which are neither horses, mules, donkeys, camels, nor elephants; for instance, the llamas used in South America, yaks in Thibet, and oxen in many parts of the world. Boats need not always be comprised under sailing-boats, steam-boats, and row-boats; thus, there are boats with paddle-wheels worked by a handle or crank inside the boat; there are also canal-boats towed by horses or men, ferry-boats moved by the force of a river, barges which go up and down with the tide in a river.

41. All these difficulties are avoided in the **perfect logical method of dividing each genus**

into two species and not more than two, so that one species possesses a particular quality, and the other does not. Thus, if I divide dwelling-houses into those which are made of brick and those which are not made of brick, I am perfectly safe, and nobody can find any fault with me. Even if I do not know what dwelling-houses are exactly, yet I may be quite sure that anything which is a dwelling-house will belong either to the species made of brick, or, if not, to the other species which are not made of brick. But this would not be the case if I divide the genus at one step into many species. Suppose, for instance, that I divide dwelling-house as below:—

The evident objection will at once be made, that houses may be built of other materials than those here specified. In Australia houses are sometimes made of the bark of gum trees; the Esquimaux live in snow houses; tents may perhaps be considered as canvas houses, and it is easy to conceive of houses made of terra-cotta, paper, straw, &c. All logical difficulties will however be avoided if I never make more than two species at each step, in the following way:—

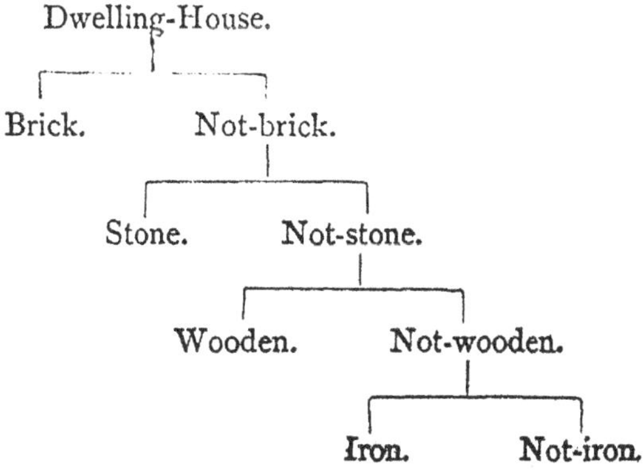

It is quite certain that I must in this division have left a place for every possible kind of house; for if a house is not made of brick, nor stone, nor wood, nor iron, it yet comes under the species at the right hand, which is not-iron, not-wooden, not-stone, and not-brick.

42. If, again, we divide substances into the two species, solid and not solid, every substance must fall into one or the other, and nothing can fall into both. No doubt there are degrees of solidity, and we might meet with substances such as tar, treacle, putty, &c., which might be said to be in a half solid state. But, if they are only half solid they must not be put into the class of solid things, and therefore they must go into the class of things which are not solid. If requisite we may make a new class of viscid things, or semi-fluid things, and we may go on, time after time, making divisions in the same way. We should get some such series of divisions as the following:—

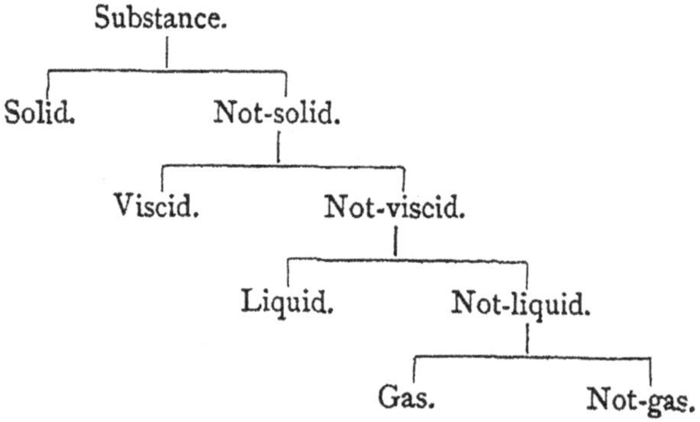

We must understand, in reading the above, that liquid things are both not viscid and not solid, and that gas is not liquid, not viscid, and not solid. No possible logical fault can be found with this; for, if we really know what we mean by a solid, viscid, liquid, and gas, any substance whatever must fall under one division and only one. If we can find any substance, such as india-rubber or jelly, which does not correspond to any of the descriptions of solid, viscid, liquid, or gas, there still remains a division provided for it, namely, that of not solid, not viscid, not liquid, not gas.

This manner of classifying things may seem to be inconvenient, but it is in reality the only truly logical way. Other methods of dividing a genus into species are only correct so far as they are constructed on the same principle, though this may not be apparent.

43. Let us inquire exactly what we do when we take brick-dwelling-house as a species of the genus dwelling-house. There are certainly not so many brick-dwelling-houses as there are dwelling-houses, because we exclude from the species all stone, wood, iron, or other kinds of dwelling-houses. Thus we find that **the species has a narrower extension than the genus** (Art. 22). In one way it has less meaning than the genus, because there are fewer objects called brick-dwelling-houses, than those which may be called dwelling-houses. But in another point of view there is more meaning in the species than in the genus, because we know more about the things. We know that anything placed in the class brick-dwelling-house is not merely a dwelling-house, but that it is made of bricks. This we may express by saying that **the species has greater intension than the genus**, meaning by intension (Art. 22) the number of qualities which belong to all things in the class.

44. The quality by which a genus is divided into two or more species is called **the difference.** In the last article, brick, or "made of brick," is the circumstance by which the species of brick-dwelling-houses is distinguished from all other dwelling-houses. Thus we may be said to add the quality "made of

brick" to the qualities of a dwelling-house, in order to get the qualities of the species we want. These qualities, namely those common to all of the genus, with the difference added, make the definition of the species. **By a definition we mean a precise statement of the qualities which are just sufficient to mark out a class,** and to tell us exactly what things belong to a class and what do not. Nothing is more important than to be able to define clearly the classes of things about which we are debating, but this is often a difficult task. In this case the definition of brick-dwelling-house, will consist of the difference, "brick," added to the definition of dwelling-house, which again might be said to consist of the circumstance that the house is used for dwelling in, added to the definition of a house.

45. We must not suppose for a moment that all the qualities of a thing are to be included in its definition. A certain quality may belong to some of a class and not to others, in which case it obviously cannot be part of the definition. Some bricks are red, some white, and some blue; the quality redness, then, will be no part of the definition of brick-dwelling-house, but will be said to be **an accident of the species.** Thus by an accident we mean any quality or circumstance which may or may not belong to a class, accidentally as it were. There are other qualities which belong to the whole of a class and yet are not regarded as part of the definition. **Such qualities are called properties of the class.** We might perhaps say that it is a property of all brick dwelling-houses to be durable. It is a property of the class mushroom to be good to eat; it is a property of all the large class of grasses to be not poisonous.

46. It will now be understood how important it is to be able to classify and define things accurately,

because when once we can do this, the properties which belong to the things will also be readily known. The qualities of the things which we meet with around us are not mixed up without order, but some of them follow from or are attached to other qualities. This is very well seen in the case of geometrical figures. We define the species triangle, as containing "three-sided rectilinear figures." The genus is rectilinear figure, or "figure made entirely of straight lines," and the difference is "three-sided," by which triangles are distinguished from figures of four, five, or more sides. But triangles besides being three-sided rectilinear figures have many other properties always present. The three angles of a triangle, when added together always make exactly two right angles. If lines be drawn through the middle of each side of a triangle perpendicularly to the side, they will all meet in one point, and so will lines drawn through the angles, and dividing them equally. There are a great many other circumstances true of all triangles, as may be learnt in any book on geometry, and all these may be rightly called properties of triangles. A circle may be defined as a plane figure, every point in the boundary of which is equally distant from a single point, but the properties of circles are exceedingly numerous, and are not fully described in any book.

VIII.—PROPOSITIONS.

47. Having now sufficiently learnt the nature and use of logical terms, we come to the second part of logic, which describes propositions. As we learned at the beginning (Art. 11), an ordinary proposition joins two terms together by means of a verb called a **copula**. It is only when we thus assert some agreement or connection between terms, or assert one

thing of another that we can be said to be right or wrong. If I were to say "The weather" without saying anything more, no one could know what I meant, or whether I meant anything at all. Nobody could answer me, or say that I was either right or wrong. But, if I say "The weather is hot" people can judge whether there is an agreement between the terms corresponding with what they feel. Let us inquire exactly what is the meaning of a proposition.

Take as an example "Coins are metallic." Here we have one concrete general term, Coins, joined to another concrete general term, metallic, which may be considered to mean "made of metal." The proposition states that the quality of being made of metal belongs to all coins. The things about which we are chiefly thinking are coins, and the term coins is therefore said to form the **subject of the proposition.** In most cases we may know the subject of a proposition by its being put first. The copula "are" comes next, and joins the subject to words indicating the quality which belongs to it, namely "metallic." This is called **the predicate of the proposition**, which is merely a word derived from the Latin, and meaning **that which is stated or affirmed.** A proposition consists, then, of subject, copula, and predicate in the order as thus stated.

48. We may explain the meaning of a proposition in another way, which, however, comes to the same thing in the end. There are great numbers of coins in the world, and still greater numbers of things made of metal. When we say "Coins are made of metal," we assert that all coins will be found among the things made of metal. If we could imagine all the metallic things in the world put into a heap together, and if we then picked the coins out of them, we should get all possible coins, because, if there were any not

in the supposed heap, they would not be made of metal, all things so made having been put into this heap. We come to this result then, that a proposition of the kind described asserts that **the subject is the name of a thing, or class of things, contained among the more numerous things of which the predicate is the name.**

49. I have said that a proposition consists of subject, copula, and predicate joined together in the order as stated. But they are not always given in this way in writing and speaking. Sometimes the proposition is inverted and the predicate is put first, as in "Blessed are the peacemakers," "Strong is truth." In such cases we must judge as well as we can which is the subject and which the predicate by the character of the words or their meanings. Thus the words "blessed" and "strong" being both adjectives are evidently predicates. Very commonly, again, the copula is not distinctly expressed but is contained in a verb. "The sun shines" seems to be a proposition with two terms and no copula, but it really means "The sun is shining." In Latin a single verb may make a complete proposition, as in "Amo," I love. When Cæsar said "Veni, Vidi, Vici," I came, I saw, I conquered, he expressed three complete propositions in three words. The science of language, however, shows that each of these single words arose from the joining together of the subject, copula, and predicate, in the same way that we shorten "I am" into "I'm," or "do not" into "don't."

50. There are, however, various kinds of propositions, and that as yet considered belongs to the **affirmative** kind. **Negative propositions,** on the contrary, **assert that the subject is not contained among the predicate.** When I say "Coins are not combustible" I think at the same time of two classes of things, "Coins" and "com-

bustible things;" but I come to the conclusion that the coins would not be found among combustible substances, such as wood, coal, oil, gas. If we had a museum which contained nothing but combustible things, there would not be a single coin shown in it. Similarly, in a museum of coins we shall not find any combustible thing shown as a coin. Thus the negative proposition in question asserts that the subject and predicate are altogether separate, and that no object belonging to the one class is found likewise in the other. We may know a negative proposition by its containing the little word "not," or it may be "no;" but sometimes such words as "never" or "nowhere" are used to make negative propositions.

51. So far it has seemed as if there were only two kinds of propositions, affirmative and negative. But, before we go on, I ought to say that propositions may be divided in a quite different way. **Hypothetical propositions** do not positively assert the predicate of the subject, except under certain circumstances. Thus, "if water be boiling, it will scald" is a hypothetical proposition asserting, not that all water will be found among scalding things, but that, when it is boiling, it will scald. "If gunpowder be damp, it will not explode;" this is a negative hypothetical proposition; for it asserts that gunpowder, when it happens to be damp, will not be found among exploding things. Hypothetical propositions may generally be recognised by containing the little word "if;" but it is doubtful whether they really differ much from the ordinary propositions already considered. We may easily say "boiling water will scald," and "damp gunpowder will not explode," thus avoiding the use of the word "if."

52. Propositions belonging to a third class are called **disjunctive**, and contain the little conjunction "or," sometimes together with "either." As examples we may say: "Lightning is sheet or forked;"

"Arches are either round or pointed;" "Angles are either obtuse, or right angled, or acute." These propositions, as we see, contain more than one predicate, and do not say to which the subject belongs. Arches are not always round, and if not round are pointed, and if not pointed they are round. There is a choice of predicates. Disjunctive propositions are very important, but more difficult to understand than other kinds of propositions, and it will be convenient to leave their further consideration until after we have learnt the nature of syllogistic reasoning.

53. We have already learnt that propositions may be affirmative or negative. They differ also as regards what is called **the quantity of the proposition,** which depends upon the quantity of the subject of which the predicate is held to be true. When I say "All clouds in the sky are composed of particles of water" I mean to assert that the whole quantity of clouds appearing high up in the atmosphere are to be found among things composed of minute particles of water. There are other things also formed of such particles, namely mists, fogs, spray, steam, &c. I may say then that the predicate in this proposition belongs universally to all clouds in the sky, and the statement is accordingly called a **universal proposition.**

54. If I say again, "Some persons are deaf-mutes," the quantity of the subject persons is said to be particular, because, as shown by the little adjective "some," I do not intend to assert that more than a portion of the subject "persons" are known to be in the class of deaf-mutes. Every proposition in which the predicate is stated to belong to a part of the subject is called **a particular proposition.** As other instances I may mention such as the following: "a few Englishmen can speak Chinese;" "many Englishmen emigrate;" "certain books are intended only for

reference;" "most storms are preceded by a fall of the barometer." Particular propositions may be either negative or affirmative; thus, "some well-water is not fit to drink" is a particular negative proposition. Universal propositions also may be either negative or affirmative, so that as twice two make four, there come to be four principal kinds of propositions, namely, universal affirmative propositions, universal negative propositions, particular affirmative propositions, and particular negative propositions. We must go on to inquire more exactly into the nature and meaning of each of these four kinds of propositions.

55. When we intend to make a statement about all the things which can be included under a term, we are said to take the term **universally**, or as logicians often say, **the term is distributed.** In the proposition "all coins are made of metal," the term "coins," as already explained, is taken universally, or is distributed, because the little adjective "all" indicates that the statement applies to any and every coin. But **the predicate is only taken particularly and is not distributed**; it would be absurd to suppose that we intended to state that all things made of metal are coins. We can only have meant that all coins are among things made of metal, or are a part of them, and there exists, of course, an almost numberless variety of other things made of metal. We must carefully remember then that **a universal affirmative proposition** like the one we have been examining, **distributes its subject, but does not distribute its predicate.**

56. We may show very clearly the exact meaning of a proposition by imagining that the things we are speaking of are included in circles, like sheep in sheep-pens. Imagine that all things made of metal and only such are put in the larger circle in Fig. 1, and all coins in the smaller circle. As the smaller

circle lies within the larger one, it follows that all coins are included among things made of metal, there being

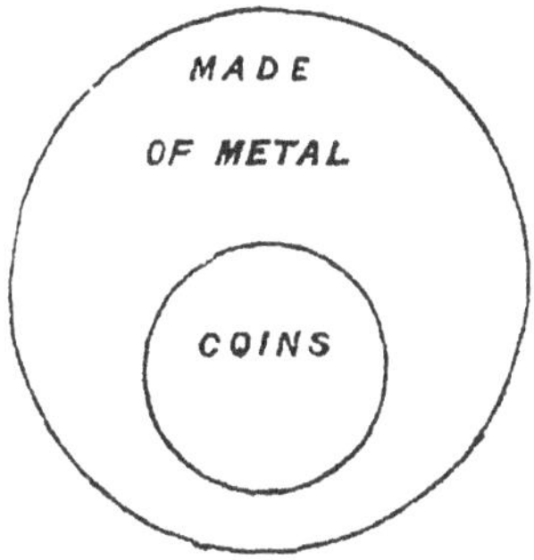

FIG. 1.

nothing but such inside the larger circle. We shall often find it convenient to use circles to show how one class or term is included wholly or partly in another, or excluded from it, as the case may be.

57. As a universal negative proposition, let us take "No sea-weed is a flowering plant," and inquire carefully what this means. It evidently speaks of all sea-weeds, so that the subject is distributed; but does it take the predicate, flowering plant, in a universal sense? Our answer should depend upon whether or not we must examine all flowering plants before we decide that no sea-weed is a flowering plant. But, if we omitted to consider a single flowering plant, and this proved to be a sea-weed, our proposition would be untrue. The proposition asserts, then, that no sea-weed is the same as any flowering plant, so that there is complete separation between the two classes, and no object can be placed in both classes.

58. We may show this in Fig. 2, the circle supposed to contain all sea-weeds lying quite outside of the circle containing all flowering plants.

If any part of one circle were to lie over part of the other, some objects would be in both classes whereas the proposition asserts that no sea-weed is in any part of the class flowering plant. We arrive,

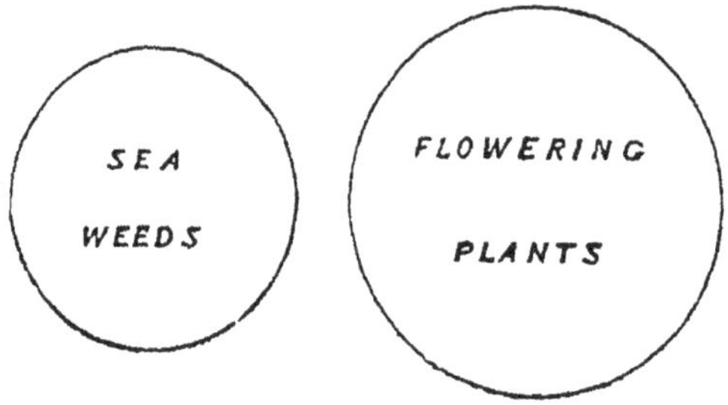

FIG. 2.

then, at this important truth, which should be carefully borne in mind, that **the universal negative proposition distributes, or takes universally, both its subject and its predicate.**

59. We shall have no difficulty in seeing that **a particular affirmative proposition distributes neither its subject nor its predicate.** Take as an example, "some violets are odorous." The subject "violets" is, of course, undistributed, because the proposition is particular. The predicate, moreover, is undistributed; for it cannot be supposed that we intended to say that some violets are the only odorous things. There are a multitude of other flowers, and many substances which are odorous in addition to violets, so that the proposition must be taken as "some violets are some odorous things," or a part of odorous things. The predicate, then, as well as the subject, is taken particularly, or is undistributed.

As other examples of the same kind of proposition I might mention the following:—many foolish novels are published; most tunes in a minor key are melancholy; a few specimens of Saxon architecture still exist; threepenny pieces are sometimes mistaken for fourpenny pieces.

60. Coming, lastly, to a particular negative proposition, say "some violets are not odorous," we know that **the subject is undistributed, but we may easily discover that the predicate is distributed.** Unless the some violets, of which we are speaking, were quite shut out of the class of odorous things, it would be untrue that they were inodorous. Hence we really mean that "some violets are not any odorous things," so that the predicate "odorous things" is taken universally.

61. When we try to show the meaning of particular propositions by using circles, it is difficult to avoid mistakes; but we often make mistakes of the same kind in thinking and talking, and it is well to be aware of the fact. When we say "some violets are odorous," we should generally be supposed to mean that "some violets" are so, and others are not; but in this case one affirmative proposition really means the same as an affirmative one and a negative one put together, namely :—

Some violets are odorous;
Some violets are not odorous.

But it is not logical to say one thing and mean another. When we say "some violets are odorous," we ought to be understood as meaning simply that "some are," leaving it quite uncertain whether other violets are or are not. In many cases we really should not know. I may safely say, for instance, that "some dogs are descended from wolves," it being nearly certain that some dogs are so; but it may be afterwards ascertained that all dogs are so descended, or, on the contrary, that some are not so. I may say again that "some metals are combustible," without meaning to say that some are not. I may correctly say that "some men or most men laugh," without staying to inquire carefully whether all men do as a fact laugh. Not being sure that some men do not laugh, I must not be supposed to assert this, in saying that some do. In the absence then of any knowledge to the contrary, **the word some must be taken to mean "some and it may be all."** I may safely say "some, and it may be all, dogs are descended from wolves," though it may afterwards be shown to be untrue that all dogs are so descended.

62. Returning to the use of circles to show the meaning of the propositions in question we meet a similar difficulty. If I draw two circles crossing each other as in Fig. 3, and fill one circle with violets and

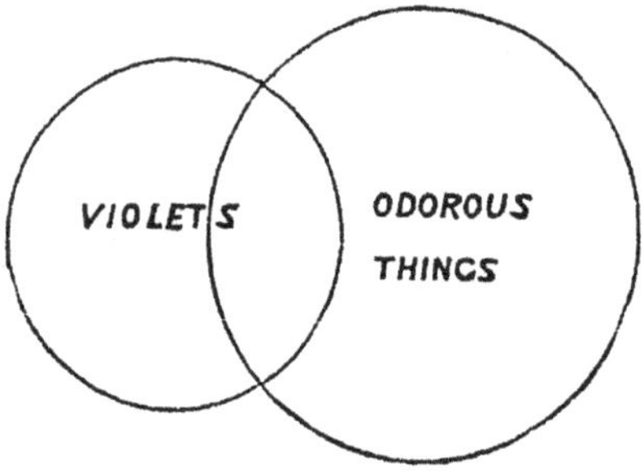

FIG. 3.

the other with odorous things, the figure evidently means that part of the class violets is in the class

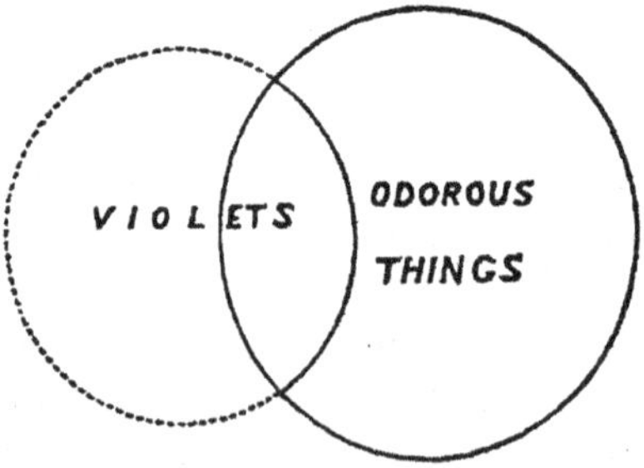

FIG. 4.

odorous things; but then another part of the same class violets is outside the odorous things, so that both the particular affirmative and the particular negative are shown at the same time. To avoid the difficulty we might perhaps use a circle with a part of its circumference broken. Thus, Fig. 4 would show that there certainly existed some violets inside the circle of odorous things, but the broken line might be understood to mean that it was doubtful whether or not any violets were really outside the odorous things. Such a figure then indicates the meaning of the particular affirmative proposition. If the broken part of one circle lies inside the other circle, as in Fig. 5, the meaning will evidently be that

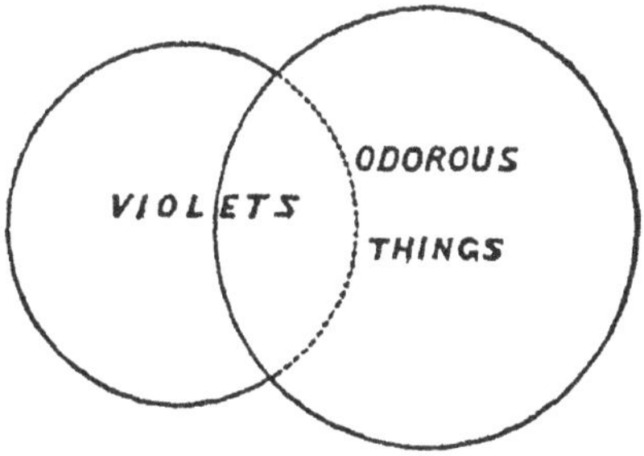

FIG 5.

some violets are known to be outside the odorous things, but that it is doubtful whether some violets are inside or not. This is the true meaning of the particular negative proposition.

IX.—HOW TO CHANGE PROPOSITIONS.

63. Having now carefully learned the nature of each of the four chief kinds of propositions, we must consider various ways in which we can draw or infer one proposition from another. We can often put the same truth into different words, just as we can mould the same clay into different forms, though it always remains the same clay. We can do likewise with

propositions; it comes to the same thing, for instance, whether I say "all coins are metallic," or "no coins are not metallic;" or again "there are no coins which are not metallic."

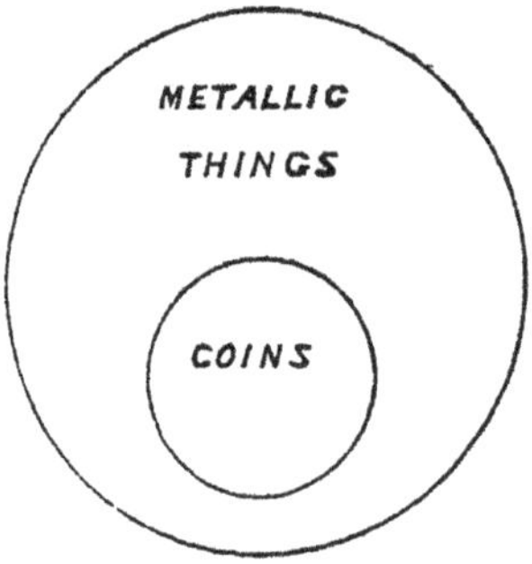

FIG. 6.

64. If, using circles again (see Fig. 6), we suppose all metallic things to fill up the larger circle, it follows that everything which is not metallic is outside the circle; and, as all coins are supposed to be within the smaller circle, included in the greater one, it follows that none of the coins can be outside the greater circle, or among non-metallic things. It is evidently the same in the end to say that all coins are within the circle of metallic things, and that none of them are outside. In this way we can always change a universal affirmative proposition into a universal negative one of the same meaning, and we can make the change backwards again. Thus, to say "there are no things which may not be useful," is only a longer way of saying, "all things may be useful." It is very desirable that the reader should practise himself in quickly and correctly making this and several other changes of propositions which I shall describe.

65. We can always change a proposition by turning it about, so as to make the old subject into a new

predicate, and the old predicate into a new subject. We are then said **to convert the proposition,** and the new proposition is called **the converse** of the old one. But it does not follow that the new one will always be true if the old one was true. Sometimes this is the case, and sometimes it is not. If I say, "some churches are wooden buildings," I may turn it about and get, "some wooden buildings are churches;" the meaning is exactly the same as before. This kind of change is called **simple conversion,** because we need do nothing but simply change the subjects and predicates in order to infer a new proposition. We see that **the particular affirmative proposition can be simply converted.** Such is the case also with the universal negative proposition. "No large flowers are green things" may be converted simply into "no green things are large flowers," by merely writing, "green things" in place of "large flowers," and large flowers instead of "green things."

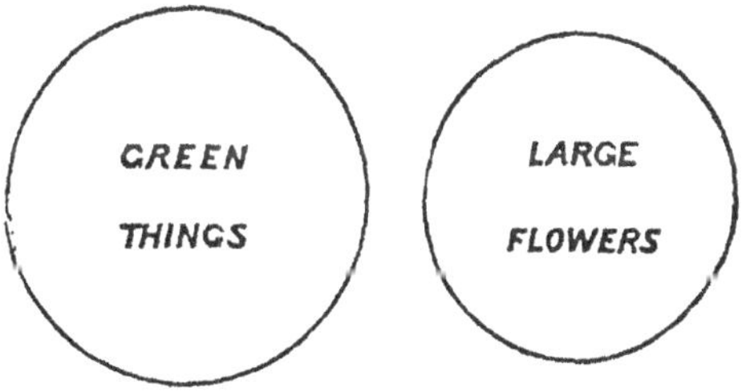

FIG. 7.

Using circles (see Fig. 7), since the green things are quite separated from the large flowers, it evidently follows that the large flowers are quite separated from the green things.

66. It is a more troublesome matter, however, to convert a universal affirmative proposition. The statement that "all jelly fish are animals," is true;

but, if we simply convert it, getting "all animals are jelly fish," the result is absurd. This is because, as we learned before (Art. 55), the predicate of a universal affirmative proposition is really particular. We do not mean to say that jelly fish are "all" the animals which exist, but only "some" of the animals. The proposition ought really to be, "all jelly fish are some animals," and if we converted this simply, we should get, "some animals are all jelly fish." But we almost always leave out the little adjectives *some* and *all* when they would occur in the predicate, so that the proposition, when converted, becomes "some animals are jelly fish." This kind of change is called **limited conversion**, and we see that **a universal affirmative proposition when so converted gives a particular affirmative one.**

67. This may seem all very plain and evident when we think about it carefully, yet it is very common to meet with people who fall into mistakes by hasty and careless thinking. By frequently seeing animals, we learn that they are all capable of moving themselves in some way, and we get so accustomed to think "all animals are moving things," that, whenever we see a thing moving of its own accord, we are inclined to infer that it is an animal. We convert the proposition wrongly, and infer that "all moving things are animals." This is quite untrue; for not only are there sensitive plants, fly-catchers, sun-dews, and some other large plants, which move almost like animals, but there is an immense number of very small plants, visible only in a good microscope, which continually move about quite as quickly as small animals. It is a curious fact, too, that very small particles of clay, mud, glass, or sand, when put into pure rain water, and examined by a strong microscope, are found to skip about as quickly as insects.

68. It is not unnatural, however, that people should

sometimes make mistakes in converting propositions of the universal affirmative kind, because in not a few cases we can properly convert them simply. This is certainly the case when the subject and predicate are singular terms (Art. 14). Thus, "The Prince of Wales is the Duke of Cornwall," and we may of course convert this simply into "The Duke of Cornwall is the Prince of Wales." The poet Pope says, "The proper study of mankind is man;" but we express exactly the same meaning if we say, "man is the proper study of mankind."

69. In other cases general terms may exactly coincide one with another. It is a truth easily proved in geometry that all triangles with three equal sides have three equal angles; at the same time, all triangles with three equal angles have three equal sides. So that we might express the two truths at once, by saying, "all triangles with three equal sides are all triangles with three equal angles." This would be converted simply into "all triangles with three equal angles are all triangles with three equal sides." Whenever we meet, then, a proposition stating that one thing or class "is" another, or agrees with another, we ought to take the trouble to ascertain exactly whether the subject agrees with or makes the whole of the predicate or only part of it. In "all jelly fish are animals," of course the jelly fish are a small part only of the animals; but the triangles with three equal sides exactly agree with the triangles with three equal angles, and there are no other triangles with three equal angles except those which have three equal sides.

If we want to put one of the propositions which we have just been considering into the form of a circular diagram, a single circle will suffice. The circle containing "man" ought exactly to cover and coincide with that of the "proper study of mankind," if the poet Pope be correct. This is shown in Fig. 8.

FIG. 8.

70. There is yet another and a rather more difficult way of converting universal affirmative propositions. If "all coins are metallic," it follows that "all not-metallic things are not coins;" but some people appear to be unable to see at first sight that this follows. A diagram, however, will make it plain. In Fig. 9, all metallic things are supposed to be inside

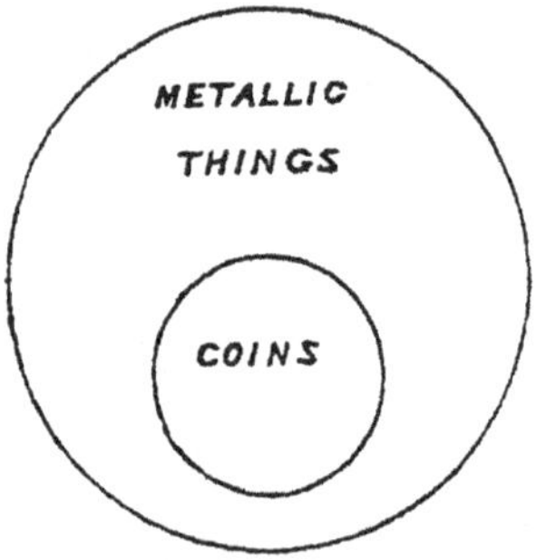

FIG. 9

the larger circle, and all not-metallic things outside this circle. Now, as all coins are within the smaller circle, it is evident that none of the not-metallic things, which are outside the larger circle can be inside the smaller circle. Or, we may explain it in this way:—If all coins are metallic, it is impossible that what is not-metallic should be a coin, for then it would be also metallic, or the same thing would be at the same time not-metallic and metallic, which is absurd. From every universal affirmative proposition we may then infer a new proposition, which has the negative of the former predicate as its subject, and the negative of the former subject as its predicate.

We can also make the same change backwards; from "all not useful beings are not living beings," we can infer, "all living beings are useful beings." For if we proceed to convert this last proposition in the way described, we get, "all not useful beings are not living beings," which is the proposition with which we began.

X.—SYLLOGISM.

71. In a great many of the arguments which we most commonly use, one proposition is gathered or inferred from two other propositions. It is well known, for instance, that, "all English silver coins are coined at Tower Hill," and it is also known that, "all sixpences are English silver coins." It follows that "all sixpences are coined at Tower Hill." These propositions are of the kind called universal affirmative, but we may give different names to them nevertheless, according to the place they hold in the reasoning. That last proposition which we gathered from the first two is called the **Conclusion**, probably because the argument is finished when we have learnt what it should be. The other two propositions, from which we gather or infer the conclusion, are called **premises**, because they are put forward, or put first, for the purpose of being reasoned about.

72. There will be no difficulty in seeing why the conclusion follows from the premises in the case given. For one premise tells us that "all English silver coins are among those coined at Tower Hill," though they are not the whole, as gold and bronze coins are also made there. The other premise informs us that "all sixpences are among English silver coins," sixpences being again a part only of such silver coins. If we take three circles to contain respectively sixpences, English silver coins, and things coined at Tower Hill, as in Fig. 10, we see that sixpences are

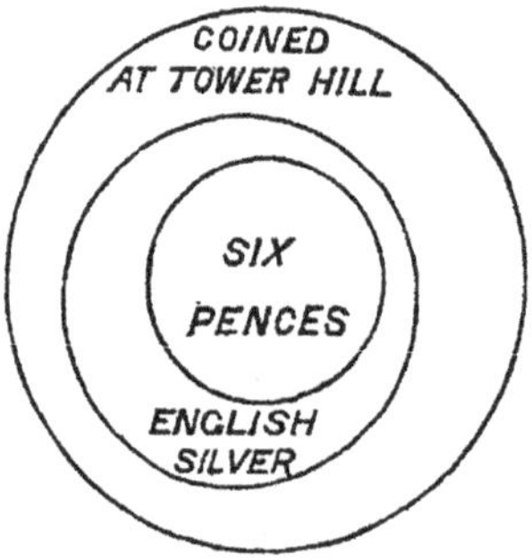

FIG. 10.

among the things coined at Tower Hill, because they are among the English silver coins, which are coined there.

73. As a second example of an argument in which we draw one proposition from two others, we will take the following:—

All electors pay rates;
No paupers pay rates;

Therefore, no paupers are electors.

Here the conclusion is a universal negative one, and it is inferred from two premises, the first of which is a universal affirmative, and the second a universal negative proposition. We may explain the reasoning in this way: all electors are among those who pay rates, whereas paupers are not among those who pay rates; therefore the paupers are quite separated from the electors. Making use of circles again, we see that the circle of electors is inside that of those who pay rates, whereas the circle of paupers is outside, so that no part of the paupers' circle can touch or overlap that of electors.

74. Although in these, and in some other cases, it is very easy to see that the conclusion will follow

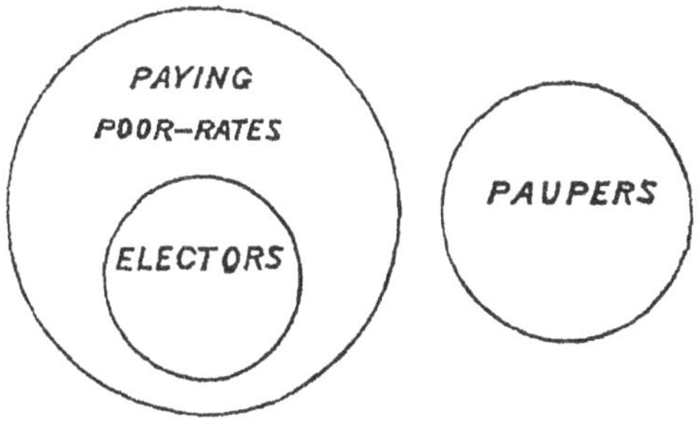

FIG. 11.

from the premises, this is not always the case. We must therefore examine how good syllogisms are made up, and what rules we must obey in making them. We will take again for this purpose our former example:—

All English silver coins are coined at Tower Hill;
All sixpences are English silver coins;

Therefore all sixpences are coined at Tower Hill.

We may observe that there are only three terms or classes of things reasoned about, namely, sixpences, English silver coins, and things coined at Tower Hill. Of these, the class of English silver coins does not occur in the conclusion; it is only used to enable us to compare or join together the other two classes of things, and in the diagram (Fig. 10, Art. 72) its circle lies between the other two circles. Accordingly, it is named **the middle term.** The largest circle is that containing all things coined at Tower Hill, the predicate of the conclusion, and this is called the **major term of the syllogism,** that is the **larger term.** Sixpences, on the contrary, being in the smallest circle, form the **minor or the lesser term,** which is always the subject of the conclusion.

75. We shall have a great deal to do with major and minor and middle terms, and therefore I must ask the learner to remember carefully that **the middle term is always the term which is not in the conclusion; that the major term is the predicate of the conclusion; and that the minor term is the subject of the conclusion.** It is also convenient to give separate names to the two premises, and that which contains the major term is always called **the major premise,** and that which contains the minor term, **the minor premise.** It is thought to be more correct to write the major premise first, but even if it be put second it is still called the major premise because it contains the major term.

XI.—THE RULES OF THE SYLLOGISM.

76. To find out whether an argument which seems to be a syllogism is really a syllogism, we must examine it carefully, and ascertain whether it agrees with certain rules. The great logician Aristotle more than two thousand years ago discovered these rules and showed how to decide when supposed syllogisms are good, and when they are not good. Several logicians have in the last fifty years been trying to

find out some simpler and better mode of ascertaining when arguments are good, but they have not yet agreed upon the subject. Until they do agree upon something better, we shall do well to learn the old rules, which are certainly both ingenious and useful.

77. *Rule I.*—In the first place **a syllogism must contain three terms, and not more than three terms**; for the reasoning consists in comparing two terms with each other by means of a third term, which we have called the middle term. If, then, there were four terms, the argument would consist either of two syllogisms, or of none at all. Suppose the terms to be cow, cloven-footed animal, ruminating animal, and animal having two stomachs. I may say that "all cows are cloven-footed animals," and that "all ruminating animals have two stomachs;" but this will not give the conclusion "all cows have two stomachs," unless we have yet another proposition comparing cloven-footed animals with ruminating animals. But, with this third proposition, we can make two complete syllogisms, the first proving that cows are ruminating animals, because they are cloven-footed, and all cloven-footed animals are ruminating animals; and the second in like manner showing that because cows are ruminating animals, therefore they have two stomachs.

A syllogism then must have just three terms, neither more nor less, and these terms are called, as we have already learned (Art. 74), the major, middle, and minor terms.

78. *Rule II.*—**A syllogism must consist of three propositions, and only three propositions,** of which one is the conclusion, and the other two are the major and minor premises. For if there be four propositions, one will be the conclusion and the other three premises. But two premises are sufficient to compare two terms with a middle term,

so that three premises will either make no such comparison at all, or will make two syllogisms. We may easily see this by considering again the case of cows. Two propositions enable us to show that a cow is a ruminating animal, because it is cloven-footed; and a third proposition enables us to make a new syllogism showing that it also has two stomachs.

79. *Rule III.*—It is an important rule that **the middle term of a syllogism must be distributed, that is, taken universally, or in its whole extent of meaning, once at least** in the premises. The reason for this rule is not quite so easy to explain, but it will afterwards be made pretty evident by examples. It amounts to this, that unless we take the whole of the middle term once, the two premises may refer to different parts of the middle term, so that there may really be no true middle term at all. If I say that "some animals are flesh-eating animals," and "some animals have two stomachs," it would be absurd to infer that therefore flesh-eating animals have two stomachs. The "some animals" which are flesh-eating, may be, and in fact are, quite distinct from the other "some animals" which have two stomachs. We may in fact say that there are four terms, and that we thus break the first rule of the syllogism, although there seem to be only three terms. But if I argue that, because "some animals are flesh-eating," and "all animals consume oxygen," therefore "some animals consuming oxygen are flesh-eating," there must be a good middle term. The "some animals" in the major premise must be part of the "all animals" in the minor premise, and thus we have a sure means of comparison between the major and minor terms.

80. *Rule IV.*—This rule is to the effect that we must not infer anything about the whole of a term,

unless something was said about the whole of the term in the premises. In other words, **no term must be distributed in the conclusion unless it was distributed in the premises.** It would be absurd to argue that because brittle substances are not fit for coining, and some metals are brittle substances, therefore no metals are fit for coining. We can, of course, infer that "some metals" are not fit for coining, namely, those which are brittle; but to include other metals as well is simply to suppose we have knowledge about them which is not given in the premises at all. It is not always so easy to find out when this rule is broken. To go back to the example in Art. 79, because some animals eat flesh, and all animals consume oxygen, we must not conclude that all which consume oxygen eat flesh. We must remember that the minor premise, "all animals consume oxygen," is an affirmative proposition, which, as fully explained in Art. 55, does not distribute its predicate, that is, does not refer to all things which consume oxygen. In other cases the way in which this fourth rule is broken will be still less apparent at first sight,

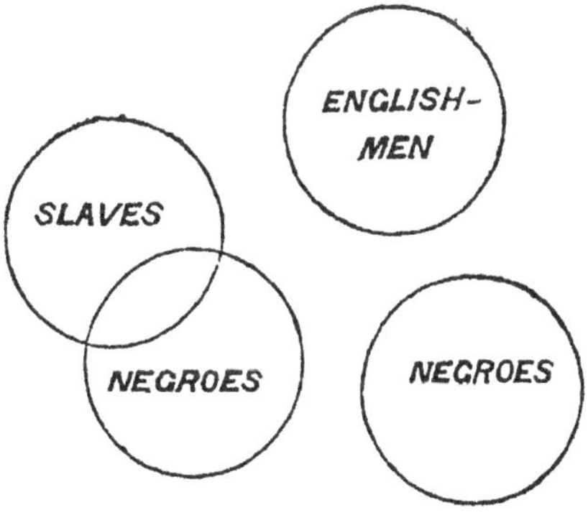

FIG. 12.

but these cases will be described further on (Art. 87, 88).

81. *Rule V.*—It is very certain that **from two negative premises nothing can be inferred.** A negative proposition asserts that two terms differ, so that the classes of things denoted by the terms are wholly or partly separated from each other. If we say then that no Englishmen are slaves, and that no negroes are Englishmen, we must represent the Englishmen by a circle quite separate from that of the slaves, and the negroes by a circle quite separate from that of Englishmen. But then we shall see after very little consideration that the negroes' circle may be placed either quite away from that of the slaves, or may be made to overlap it more or less. This means that negroes may be not slaves at all, or may be partly slaves and partly not slaves, or may be all slaves, for anything which the two premises tell us about the matter.

82. *Rule VI.*—The last of the principal rules of the syllogism is that, **if one premise be negative, the conclusion must be negative, and we cannot get a negative conclusion unless one of the premises be negative.** We may perhaps see the truth of this rule most easily by reflecting that a negative proposition is represented by one circle

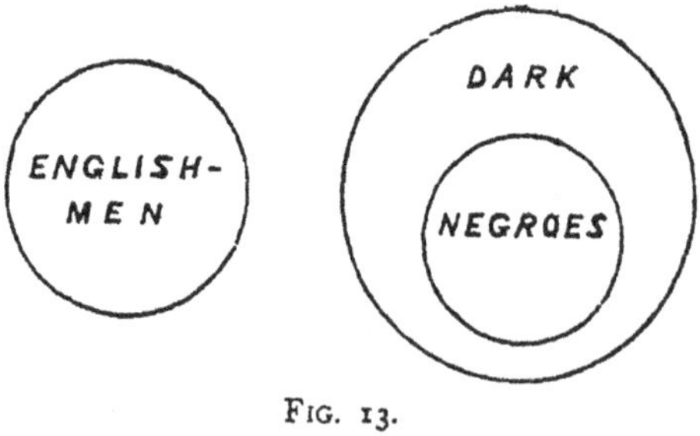

FIG. 13.

outside another. Now, if we say all negroes are dark, no Englishmen are dark, the circle of negroes is inside that of dark men, while that of Englishmen is outside, so that the circle of Englishmen must be outside that of negroes, giving a negative result. It is true that we might have the terms differently arranged. The premises might be all negroes are dark, no Chinese are negroes. The circle of negroes

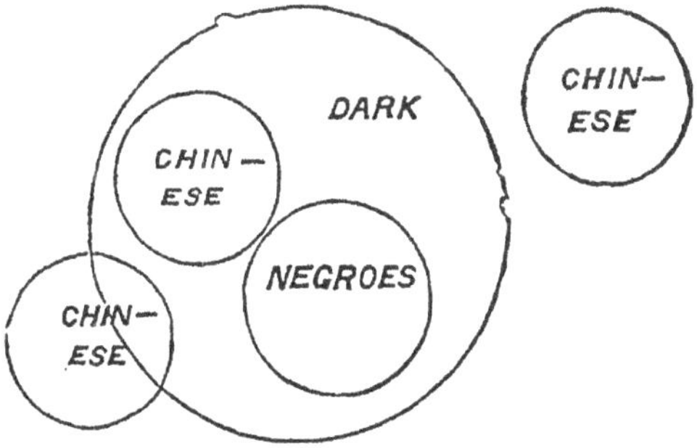

FIG. 14.

is as before inside that of dark men; but the circle of Chinese, though outside that of negroes, may be wholly inside that of dark men, or partly inside and partly outside, or wholly outside. Such premises then tell us nothing about the relative position of Chinese and dark men; we see that with one negative premise we either get a negative conclusion, or no conclusion at all.

83. A second part of the rule is that we cannot get a negative conclusion unless one premise be negative. We may satisfy ourselves that this is true by trying with circles how we can prove one circle to be outside of another by means of a third circle. This can only be done by putting one inside and one outside the third circle, and to put one outside another indicates, as we have often seen, a negative proposition.

84. Everyone who wishes to be a good logician must remember the rules of the syllogism which have now been described, and must by practice become

quick in seeing whether an argument supposed to be a syllogism does or does not obey the rules. I will give a few more examples of the way in which we must examine arguments in order to decide whether they are good syllogisms or not. Do the following premises, for instance, allow of the conclusion drawn from them?—

Every city contains a cathedral.
Liverpool does not contain a cathedral,

Therefore, Liverpool is not a city.

Here the middle term, or that which does not appear in the conclusion, is "contain (or containing) a cathedral." The minor term is Liverpool, and the major term city. There are thus three terms and no more, in accordance with the first rule, and there are three propositions and no more, in accordance with the second rule. The third rule requires that the middle-term shall be distributed, or taken universally, once at least; and this is the case, because the second premise "Liverpool does not contain a cathedral" is a negative proposition, and therefore distributes its predicate (Art. 57). As to the fourth rule, Liverpool and city are both distributed in the conclusion, but they are also both distributed in the premises, so that the rule is obeyed. The first premise is affirmative, so that the fifth rule about two negative premises cannot be broken. The sixth rule is likewise obeyed, which requires that if one premise be negative the conclusion shall be so, and this is the case. Thus, the argument we are discussing is a perfectly good syllogism.

85. Let us next examine whether the following propositions make a syllogism:—

All minerals are raised from mines.
All coals are raised from mines.

Therefore, all coals are minerals.

The middle term, which we should generally look for first, is "raised from mines;" but we ought to notice at once that both the propositions in which it appears are affirmative. Now affirmative propositions never distribute their predicates (Arts. 55, 59), so that the third rule of the syllogism is broken, which requires that the middle term shall be distributed once at least. In this case there is said to be a **fallacy of an undistributed middle term.**

86. This was the kind of fallacy into which an authoress fell when she wrote a book proving, among other things, that to wear false hair was to tell a falsehood. In reality her reasoning came to this, that to wear false hair was to deceive, and to tell a falsehood was also to deceive. But the predicate to deceive is in both cases particular and ought to be explained as meaning one way of deceiving. Now falsehood is the name for deceit by words, and is not the proper name for deceit by other means.

To make a good argument out of this matter we ought to be able to put it in this way:—

To deceive is always to tell a falsehood.
To wear false hair is to deceive.

Therefore, to wear false hair is to tell a falsehood.

This is a perfectly good syllogism supposing it to mean that every case of deceiving is a case of telling a falsehood, and if this were true the conclusion would be true. But it is evident that in the ordinary use of the word falsehood the first premise is not true. There was one philosopher who tried to prove in like manner that whenever a person did a wrong act it was only a particular way of telling a lie, so that one who killed a fellow-creature only took a round-about way of saying that he was not a fellow-creature.

87. It is not unnatural that people, who spend their

whole lives in some kind of study, should learn to perceive all its value, while, being ignorant of other branches of learning, they cannot so readily know the value of those branches. Hence they are likely to fall into the fallacy of arguing that because their own studies are very useful other studies are not. Let us take the study of Latin and Greek as an instance, and compare it with that of physical science. The argument would be put in this form :—

The study of Greek and Latin is very useful;
The study of physical science is not the study of Greek and Latin;
Therefore, the study of physical science is not very useful.

In this argument the numbers of terms and propositions are quite correct, and at the first moment it may not be easy to see where it fails. The middle term, or that which does not appear in the conclusion, is "the study of Greek and Latin." It is certainly distributed in the second premise which is negative, and may also be said to be distributed in the first premise, being in fact a singular term. One premise is negative and the conclusion is negative. So far all is right; but on making further examination, we shall find that the conclusion, being negative, distributes its predicate "very useful," while the first premise, of which it is also the predicate, does not distribute it. Thus the supposed argument breaks the fourth rule, that no term shall be distributed in the conclusion unless it were distributed in one of the premises.

88. The fact is, of course, that there may be a great many very useful studies, and because the classical studies of Greek and Latin are some of these, it does not follow that other ones are shut out. We may show this most clearly by a diagram (Fig. 15),

placing the several studies in smaller circles enclosed in the larger one of very useful studies. The circle

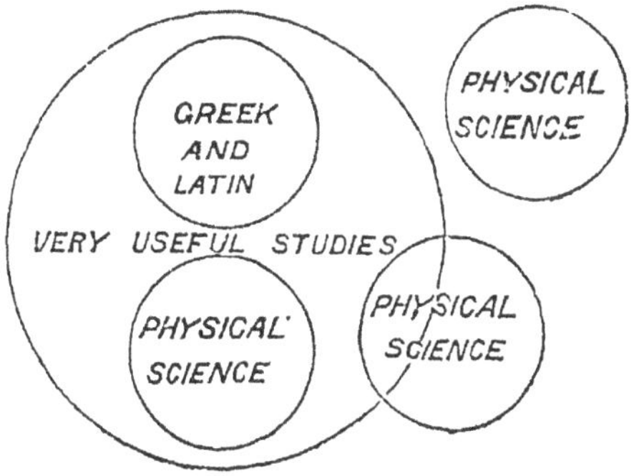

FIG. 15

of Greek and Latin must be distinct from that of physical science, and these circles must not overlap each other at all; but we see that the circle of physical science may nevertheless be placed so as to be wholly within that of "very useful studies," or partly within and partly without, or wholly without. In short, from the statement that Greek and Latin are very useful subjects of study, we get no information at all as to whether the physical sciences are or are not so. We may say the same of the study of mathematical, logical, moral, and other sciences. None of them must be considered useless, because the others are useful.

89. Suppose I were to argue that all householders pay poor rates, and all electors are those who pay poor rates; therefore, all householders are electors. Now, as a matter of fact, it is true according to the present law that all householders, excepting paupers, are electors; but does this follow from the propositions used as premises to reason upon? The middle term seems to be "paying poor rates," and

it is the predicate of both the premises, which are affirmative. Therefore it would in each case be undistributed, and by the third rule of the syllogism, the argument would be bad. But great care is often required in examining arguments, and in reality the second proposition is not what we took it to be. We do not simply say, "all electors pay poor rates," or are "among those who pay poor rates;" but we say that they "are" those, so that there are no electors (in ordinary cases) except those who pay poor rates. This is one of those propositions (Art. 68) which we can convert simply, so that we may state it as, "all who pay poor rates are (all) electors;" and as all householders pay poor rates, excepting paupers, it follows by a good syllogism, that all householders are electors.

90. There are two minor rules of the syllogism which we may deduce from the rules already given. The first is that from two particular propositions, whether affirmative or negative, we cannot deduce any logical conclusion. Thus, if we were to argue that some who elect members of Parliament are well-educated men, and some well-educated men are perfectly acquainted with what the country needs, we could not properly infer that some who elect members of Parliament are perfectly acquainted with what the country needs. The middle term is "well-educated men," and it is the predicate of the first of the propositions, so that it is undistributed. It is also undistributed as the subject of the second proposition, and thus what seems to be an argument breaks the third rule of the syllogism. As we may explain it, the well-educated men who elect members of Parliament might happen not to be those perfectly acquainted with what the country needs. In the same way, if we were to take other examples of arguments containing two particular propositions, we should find

that they can never give a conclusion according to the rules of the syllogism.

91. A second rule which follows from those of the syllogism is that, if either premise be particular, the conclusion must also be particular. If we were to argue that some electors are not fit to choose good representatives, but all well-educated men are fit to choose good representatives, therefore no electors are well-educated men, we should break the fourth rule of the syllogism. We must not infer anything at all about all electors, when in the first proposition we speak only of some electors. In a similar way every syllogism in which one premise is particular and the conclusion is not particular will be found to break one rule or other of those given in Arts. 77—82.

92. It is shown in almost all books on logic that, when we try in how many different ways we can make syllogisms with each of the four kinds of propositions variously put together, we get altogether nineteen good kinds of arguments, called the **nineteen moods of the syllogism.** These are divided into four figures, each figure being known by the position of the middle term in the premises. Logicians long ago ascertained in what cases of each figure a syllogism is valid, and they recorded the results in certain curious lines, beginning Barbara, Celarent, &c., which were so constructed that the vowels in each word show what kinds of propositions put together in a particular way will make a good syllogism. But it is not of much advantage to know these lines by heart, because we ought to understand the rules of the syllogism so well as to be able to tell in every case whether an argument is a correct syllogism or not.

93. Although every argument which is a good syllogism must consist of two premises and a conclusion, these three propositions will not usually be

stated at full length. People sometimes think that they are not arguing by syllogisms, because the parts of the syllogisms are not written or printed exactly as they are in books on logic. But they might as reasonably say that mental arithmetic is not arithmetic at all, because the sums are not worked out at full length on paper. It is not usual to state more than one premise of a syllogism in addition to the conclusion, because the reader can then judge, without much difficulty, what the other premise is intended to be. Thus in the Sermon on the Mount, the verses known as the Beatitudes consist each of one premise and a conclusion, and the conclusion is put first. "Blessed are the merciful: for they shall obtain mercy." The subject and predicate of the conclusion are here inverted (Art. 69), so that the proposition is really "The merciful are blessed." It is evidently understood that "All who shall obtain mercy are blessed," so that the syllogism, when stated at full length, becomes :—

All who shall obtain mercy are blessed ;
All who are merciful shall obtain mercy ;

Therefore, all who are merciful are blessed.

This is a perfectly good syllogism, similar to those described in Arts. 10 and 74.

94. Wherever any one of the words, **because, for, therefore, since,** or other words used in the same sense occur, we may be sure that there is an argument, and in many cases this will be found to be a syllogism. It is true that a great many of the arguments which we commonly use belong rather to geometrical or arithmetical reasoning, than to simple logic. If I were to argue, for instance, that the rocks called red sandstone lie above the coal measures, because they lie above the Permian rocks, which lie above the coal measures, this is perfectly

good reasoning. But it is not merely logical, because it deals with the position of the beds of rocks. It is a question of height, and belongs to geometry.

XII.—HYPOTHETICAL SYLLOGISMS.

95. It was stated (Art. 51) that there are supposed to be three kinds of propositions, of which the first and most common kind is employed in the syllogisms already described. We must not overlook **hypothetical propositions** which affirm something provided or "if" something else is true. By joining one such proposition with an ordinary proposition we can make a syllogism. "If Manchester contains a cathedral it is a city; but Manchester does contain a cathedral; therefore, it is a city." This is an affirmative hypothetical syllogism, and it has two premises and a conclusion, like an ordinary syllogism. The first premise is hypothetical and consists of two parts, the **antecedent** containing the little word "if," and the **consequent** which informs us what will happen under the supposed circumstances.

96. The rules of this kind of syllogism are very simple: **If the antecedent be affirmed, the consequent may be affirmed. If the consequent be denied, the antecedent may be denied.** In the instance already given the first rule applies; for we affirm that Manchester does contain a cathedral, and then affirm the consequence, that it is a city. As an example of the second rule, we may say, "If the atmosphere were equally dense at all heights there could be no perpetual snow on the Alps; but there is perpetual snow on the Alps: therefore, the atmosphere is not equally dense." This is a negative hypothetical syllogism.

97. We must take much care not to fall into **the fallacies of affirming the consequent, or**

denying the antecedent, and imagining that we are making a good syllogism. It would be wrong to argue that, "If a man is a good teacher, he thoroughly understands his subject; but John Jones thoroughly understands his subject; therefore, he is a good teacher." The conclusion may happen to be true, as a matter of fact, but it does not follow from the premises. Nor can we argue that, "If snow is mixed with salt it melts; the snow on the ground is not mixed with salt; therefore it does not melt." This argument is obviously absurd, because snow melts when warmed, as well as when mixed with salt, and by denying the one possible antecedent we leave other possible ones untouched.

98. In reality, however, hypothetical propositions and syllogisms are not different from those which we have more fully considered. It is all **a matter of the convenience of stating the propositions.** Thus, our former example (Art. 95) may be stated thus:—"All towns containing cathedrals are cities; Manchester is a town containing a cathedral; therefore, Manchester is a city." This is a good syllogism of a very common kind, the middle term being "town containing a cathedral." Our second example is not so conveniently stated as a common syllogism, but we may say, "An equally dense atmosphere is not an atmosphere allowing perpetual snow on the Alps; but our atmosphere is one allowing perpetual snow on the Alps: therefore, our atmosphere is not an equally dense atmosphere." This is a good syllogism with a negative major premise and a negative conclusion, and all the other hypothetical syllogisms can be turned into ordinary ones in the way shown by one example or the other.

99. We can now see that **to affirm the consequent and then to infer that we can affirm the antecedent, is as bad as breaking the**

third rule of the syllogism, and allowing an undistributed middle term. This is very evident in the example given (Art. 97) which becomes, "A good teacher thoroughly understands his subject; John Jones thoroughly understands his subject; therefore, John Jones is a good teacher." Both the premises being affirmative and having the middle term "thoroughly understands his subject" for their predicate, it follows that the middle term is not distributed in either premise.

To deny the antecedent is really to break the fourth rule of the syllogism, and to take a term as distributed in the conclusion which was not so in the premise. Instead of saying, "If snow is mixed with salt it melts" we may say more simply, "Snow mixed with salt melts; but the snow on the ground is not mixed with salt; therefore, it does not melt." Here the conclusion is negative, and therefore distributes its predicate "melts" or "melting." But this term occurs as the predicate of the first premise, which is affirmative, so that it is not distributed, breaking the fourth rule of the syllogism. This example is exactly like that given in Article 87.

XIII.—OTHER KINDS OF ARGUMENTS.

100. It would be quite a mistake to suppose that all good logical arguments must obey the rules of the syllogism, which we have been considering. Only those arguments which connect two terms together by means of a middle term, and are therefore syllogisms, need obey these rules. A great many of the arguments which we daily use are of this nature; but there are a great many other kinds of arguments, some of which have never been understood by logicians until recent years.

101. One important kind of argument is known as

the **disjunctive syllogism**, though it does not obey the rules of the syllogism, or in any way resemble syllogisms. We learned (Art. 52) that disjunctive propositions are those which have several terms joined together by the little word "or." We use such propositions when we divide up a class into smaller classes; thus we may say, speaking without scientific accuracy, that a vegetable is either a tree, or a shrub, or a herb. A boat is either a sailing-boat, or a row-boat, or a steam-boat. The metal of which money is made is either gold, or silver, or copper, or bronze, or nickel. There may be any number of things thus stated; for instance, a member of the House of Commons must be either Mr. Disraeli, or Mr. Gladstone, or Mr. Forster, or Sir Stafford Northcote, or any one of about 650 other men who belong to the House. Each of the things or smaller classes thus joined together by "or" will be called **alternatives,** because we may take our choice between them, and if one will not do another may do.

102. The principal rule according to which we use disjunctive propositions in arguments is that **if one or more alternatives be denied the rest may be affirmed.** Thus fuel consists of carbon or hydrogen. If then any particular portion of fuel does not consist of hydrogen, it must consist of carbon. Here there are only two alternatives, and in this and a great many like cases, if we deny one alternative we must affirm the only remaining one. A crime is either treason, or felony, or misdemeanour. Forgery is not treason nor misdemeanour; therefore, it is felony. Here we have three alternatives, two of which are denied, so that the other one alone remains to be affirmed. Roofing materials are either slates, or thatch, or shingles, or iron, or tiles, or felt, or paper. Here we have seven alternatives, and, if we held them to be all the existing ones, it would follow that a house

not roofed with slates or thatch must be roofed with shingles, or iron, or tiles, or felt, or paper. These disjunctive arguments, it will be seen, may be very various in the number of alternatives denied and affirmed; but they none of them obey the rules of the syllogism, because one proposition is always negative and yet the conclusion is affirmative, which is against the sixth rule (Art. 82).

103. It is said in some books on logic that, if we affirm one alternative of a disjunctive proposition, we must deny the remainder. It would be said, for instance, that as fuel is composed of carbon or of hydrogen, what fuel is composed of carbon is not composed of hydrogen. But this is not true, because nearly all fuel is composed of both substances at the same time. Again, it might be inferred that, as boats are either sailing-boats, or row-boats, or steam-boats, therefore a boat which is a steam-boat is not a sailing-boat, nor a row-boat. But this need not be so, and most steam-boats are able to set sails, when it is desirable or necessary to do so. A magistrate is a justice of the peace, or a mayor, or a stipendiary magistrate; but it does not follow that one who is a justice of the peace is not a mayor. After affirming one alternative we can only deny the others if there be such a difference between them that they could not be true at the same time.

XIV.—THE GREAT RULE OF INFERENCE.

104. There is a simple rule which will enable us to test the truth of a great many arguments, even of many which do not come under any of the rules commonly given in books on logic. This rule is that **whatever is true of one term is true of any term which is stated to be the same in meaning as that term.** In other words, we may always

substitute one term for another if we know that they refer to exactly the same things. There is no doubt that a horse is some animal, and therefore the head of a horse is the head of some animal. This argument cannot be brought under the rules of the syllogism, because it contains four different logical terms in two propositions, namely, horse, some animal, head of horse, head of some animal. But it easily comes under the rule which I have given, because we have simply to put "some animal" instead of "a horse." A very great number of arguments may be explained in this way. Gold is a metal; therefore, a piece of gold is a piece of metal. A negro is a fellow creature; therefore, he who strikes a negro, strikes a fellow creature. A domestic animal is a creature capable of suffering; therefore, he who ill-treats a domestic animal, ill-treats a creature capable of suffering.

105. Let it be carefully remarked that in an ordinary universal affirmative proposition, like, "A negro is a fellow creature," we cannot put negro simply for fellow creature. It would be absurd to argue that, because a man strikes a fellow creature, therefore he strikes a negro. This is evidently because negroes form only a part of our fellow creatures. But in other cases, as already mentioned (Art. 69), the subject and predicate of a proposition refer to exactly the same numbers of objects, and altogether coincide. All parallelograms, for instance, are all plane four-sided figures, whose opposite angles are equal. It follows that whatever we know of a four-sided figure of this description is true of a parallelogram, and whatever we know of parallelograms is true of such figures. Any figure which has not its opposite angles equal cannot be a parallelogram. When the terms of a proposition are singular ones, this is still more evident. The moon is the earth's satellite; it follows that any-

thing which is true of the earth's satellite is true of the moon; and anything which is true of the moon is true of the earth's satellite. The moon, as far as we can learn, is without an atmosphere, and without seas; therefore the earth's satellite is without an atmosphere and without seas.

106. It is really in the same way that we argue about quantities. Thus the length of Westminster Abbey is 505 feet; therefore, anything true of 505 feet is true of the length of Westminster Abbey. The length of Canterbury Cathedral is greater than 505 feet by 9 feet; therefore it is greater than that of Westminster Abbey by 9 feet. The width of Bristol Cathedral is equal to that of Bath Abbey Church. Hence it follows that, in respect to width, we can always put Bristol Cathedral for the Bath Abbey Church, or the latter for the former. It happens, for instance, that the width of St. Mary's Church, at Redcliffe, Bristol, is less than that of the Cathedral; hence it follows that it is less than that of the Bath Abbey Church. On the other hand, Exeter Cathedral has by accident the same width as Bristol Cathedral; therefore, putting the Bath Abbey Church for Bristol Cathedral, we find that the Cathedral of Exeter and the Bath Abbey Church have the same width.

107. When we examine carefully enough the way in which we reason, it will be found **in every case to consist in putting one thing or term in place of another, to which we know it to have an exact resemblance in some respect.** We use the likeness as a kind of bridge, which leads us from a knowledge of one thing to a knowledge of another; thus **the true principle of reasoning may be called the substitution of similars, or the passing from like to like.** We infer the character of one thing from the character of something which acts as a go-between, or third term.

When we are certain there is an exact likeness, our inference is certain; when we only believe that there probably is, or guess that there is, then our inferences are only probable, not certain.

XV.—INDUCTIVE REASONING.

108. In all the preceding parts of this Primer we have been inquiring how we may gather the truth contained in some propositions, called Premises, and put it into another proposition, called the Conclusion. We have not yet undertaken to find out how we can learn what propositions really are true, but only **what propositions are true when other ones are true.** All the acts of reasoning yet considered would be called **deductive, because we deduce, or lead down the truth from premises to conclusion.** It is an exceedingly important thing to understand deductive inference correctly, but it might seem to be still more important to understand **inductive inference**, by which we gather the truth of general propositions from facts observed as happening in the world around us.

109. It ought to be easy to see that reasoning alone will never teach us anything, because it only gives us one proposition, when we already have other ones. How then are we to get the original propositions? This must be done by using our eyes and ears, and observing things about us, so as to learn what they really are. How are we to know that all very small particles of water in daylight appear white, except by examining the appearance of clouds, mist, foam, spray, steam, and any other things which we know to be composed of small particles of water? This seems to be evidently the proper way to get knowledge, and we may well wonder that people ever thought differently. Nevertheless, for many centuries it was believed

to be possible to arrive at all necessary knowledge by the use of the syllogism, and men preferred trusting to Aristotle, rather than using their own eyes.

110. The rise of modern science may perhaps be considered to date as far back as the time of Roger Bacon, the wonderful monk and philosopher of Oxford, who lived between the years 1214 and 1292. He was probably the first in the middle ages to assert that we must learn science by observing and experimenting on the things around us, and he himself made many remarkable discoveries. Galileo, however, who lived more than 300 years later (1564 to 1642), was the greatest of several great men, who in Italy, France, Germany, or England, began by degrees to show how many important truths could be discovered by well-directed observation. Before the time of Galileo, learned men believed that large bodies fall more rapidly towards the earth than small ones, because Aristotle said so. But Galileo, going to the top of the Leaning Tower of Pisa, let fall two unequal stones, and proved to some friends, whom he had brought there to see his experiment, that Aristotle was in error. **It is Galileo's spirit of going direct to Nature, and verifying our opinions and theories by experiment, that has led to all the great discoveries of modern science.**

111. People very commonly believe that Francis Bacon, usually called Lord Bacon, who lived between the years 1561 and 1629, was the founder of inductive logic and of true scientific method. It is quite certain that Lord Bacon was an exceedingly clever man, and in many ways a great man. In his celebrated work, the **Novum Organum**, or the **New Instrument**, he strongly points out the need of observing Nature and collecting a great many facts, from which general laws might gradually be collected, and he foresaw that valuable discoveries would be made. But it is quite a

mistake to suppose that Lord Bacon really understood the inductive logic by which Galileo, about the same time, and Sir Isaac Newton and other great men after him, succeeded in detecting the chief laws of nature. Not only was Lord Bacon unable to make any real discoveries by his own methods of inquiry, when he tried to do so, but he could not see the truth of the excellent discoveries in astronomy and magnetism, which Copernicus, and an Englishman named Gilbert, had made known a little time before. Thus it is wrong to speak of Lord Bacon's philosophy as if his book the *Novum Organum* really taught men how to investigate nature, and if we continue to speak of Bacon's Philosophy, meaning the new inductive logic, **we ought to attribute it to Roger Bacon rather than to Lord Bacon.**

112. **Inductive logic inquires by what manner of reasoning we can gather the laws of nature from the facts and events observed.** Such reasoning is called **induction**, or inductive inquiry, and, as it has actually been practised by all the greatest discoverers in science, it consists in four steps.

113. In the first place, we must gain, by almost accidental observations and experiments, a knowledge of facts touching the subject of inquiry. Such knowledge of mere facts is not properly called science at all, because the facts are disconnected, and do not enable us to explain other facts, or to discover what will happen before we have tried the experiment. It is merely **knowledge given by the senses.**

114. In taking the second step, we proceed to reason about these facts, which we do by inventing or imagining laws, which may be true of the things examined. We make what is called an **hypothesis** and suppose some law or general proposition to be true for the sake of argument. We see now why

deductive logic is so very important, because it is only by deductive reasoning that we can tell what will be the consequences of the law or proposition supposed.

115. In the third step, then, we reason by the syllogism, or by other kinds of deductive argument, to the particular facts which will be true if the hypothesis be true.

116. In the fourth step, we proceed to compare these deductions with the facts already collected, or, when necessary and practicable, we make new observations and plan new experiments, so as to find out whether the hypothesis agrees with nature. If we meet with several distinct disagreements between our deductions and our observations, it will become likely that the hypothesis is wrong, and we must then invent a new one. In order to produce agreement it will sometimes be enough to change the hypothesis in a small degree.

117. When we get hold of an hypothesis which seems to give results agreeing with a few facts, we must not at once assume that it is certainly correct. We must go on making other deductions from it under various circumstances, and, whenever it is possible, we ought to verify these results, that is compare them with facts observed through the senses. When an hypothesis is shown in this way to be true in a great many of its results, especially when it enables us to predict what we should never otherwise have believed or discovered, it becomes almost certain that the hypothesis itself is a true one.

118. Thus there may be said to be **four different steps in inductive reasoning** :—

First Step.—**Preliminary observation.**
Second Step.—**The making of hypotheses.**
Third Step.—**Deductive reasoning.**
Fourth Step.—**Verification.**

I will now proceed to show by examples that it is really by this mode of reasoning in four successive steps that we learn the nature of things, and thus become able to make true general propositions about them.

119. Hundreds of years ago people had frequently noticed in stones and on the face of exposed rocks, peculiar forms closely resembling those of living animals, shells, or plants. These fossils were so remarkable that, though observed by mere accident, people could not help forming hypotheses to explain the resemblance to living beings, and very different these hypotheses were. The favourite one was that the Great Deluge carried shells, drowned animals, and other things about, and in retreating left them scattered over the surface of the earth, even upon the tops of high mountains. The celebrated Voltaire, on the contrary, suggested that the shells found high up in the Alps must have been dropped by the pilgrims, who used to cross the mountains in former centuries. Perhaps a more reasonable hypothesis was to the effect that they were "freaks of nature," that is, that the resemblance to animals and plants arose from accident, just as frost on a window-pane sometimes resembles the branches of a tree. A further hypothesis was that the fossils really consisted of the remains of living beings covered up in the mud or sand which became the substance of rocks innumerable centuries ago. The last hypothesis was selected as the true one by the processes of deductive reasoning and verification, which I have described.

120. We proceed to reason about the hypotheses somewhat in this way. If the Great Deluge deposited the fossils on mountains, then the fossils ought to be found only on the surface or near it, whereas great numbers of fossils are found in deep mines, driven through hard rocks, where the waters of the Deluge

cannot have placed them. This hypothesis, therefore, is wrong. Nor is that of Voltaire any better; for fossils are found on mountains, and in parts of the earth, the Arctic Regions for instance, where pilgrims never went, not to speak of the fossils sunk deep in the earth. The hypothesis about "freaks of nature" is less easy to disprove, and there is no doubt that at various times, things have been believed to be fossil remains of animals and plants which were not so. But we may argue in this way: if, in such a great multitude of cases, stones have been formed by mere accident in the shapes of living things, there is equal reason why they should take by accident the forms of other objects. Why should we not meet with fossil books, and fossil teapots, and fossil chairs and tables? The hypothesis of freaks of nature does not give any reason to expect what we do find, more than multitudes of things which we do not find.

121. The last hypothesis, on the contrary, namely, that an immense number of animals and plants have lived in past ages, and left their remains buried in the strata of sand and mud then deposited in the seas, lakes, or rivers, enables us to explain many peculiar facts. We see how it is possible that these remains should be found at great depths in the crust of the earth, one layer of rock after another having been formed during many millions of years. We can argue in this way too: if an animal be buried in the earth at the present day, we know that the flesh and soft parts will quickly disappear, and after the lapse of a hundred years only the bones, teeth, and hard parts will remain. Accordingly, if animals with skeletons lived in former geological ages, we ought usually to find only the bones and durable parts. And it is a fact that we possess the fossil skeletons of multitudes of animals whose forms are otherwise unknown to us. We meet too with the shells of shell-fish, the hard

scales of fishes or reptiles, the bark of trees, in short just those parts which are most durable. Sometimes even the bones of an animal have been wholly rotted away, and yet the teeth, which are the hardest and most indestructible parts of the whole body, remain.

122. We can argue, again, that if shell-fish were embedded in mud and then pressed with an immense weight of rock gradually formed over them, they ought to be compressed and flattened. Accordingly we do find fossil shells sometimes quite flat and broken as if by pressure, and the remains of the trunks of trees discovered in coal mines are never quite round, but partially flattened. In these and many other ways, then, we can argue that **if animals and plants did live millions of years ago, their remains would now present appearances which agree with what is observed.** Hence we are obliged to reject all the previous hypotheses, which disagreed with facts, and adopt the last hypothesis which so well agrees.

123. Probably the most important law of nature ever discovered is that called the **Law of Gravity,** which states that all bodies in space tend to fall towards each other with a certain force depending on the magnitudes of the bodies and the distance between them. It might seem that we need no aid of logic to show us that things fall towards the earth, because, whether we throw up a stone or a book, a gold coin or a feather, they will all descend more or less quickly to the surface of the earth. The ancient Greeks observed this much, and no doubt the ancient Egyptians and other peoples before them. But then it does not seem to be true that all bodies fall; for flames ascend upwards, and in smoke, and clouds, and bubbles we have other exceptions. Aristotle, the greatest of Greek philosophers, came to the conclusion that some things were naturally heavy and tended to

fall, while other things were naturally light, and tended to rise. Only about two hundred years ago did Newton succeed in showing how much better it was to make the hypothesis that all things tend to fall, because he could then explain not only the motions of flame and other apparently light things, but also the movements of the moon, sun, and planets. If we put a pound weight into one scale of a balance, and only half-a-pound into the other scale, the latter will of course go up as the former is pulled down by the greater force. So, if flame be a lighter substance than the air around, it will be forced or buoyed up like a cork in water. Thus, when we argue deductively, we find that what is apparently tending to rise upwards may really be tending to fall downwards, but is overpowered by the greater tendency of other bodies.

124. Newton argued again in this way: if all bodies tend to fall towards each other, all bodies ought to fall towards the earth. Now the moon is a body, and therefore it ought, according to evident reasoning in the manner of the syllogism, to fall towards the earth. Why does it not do so, but go on revolving round the earth once in every lunar month? It occurred to him that, if the moon were not in some way held by the earth, it ought to go off flying away in a straight line like a stone from a rapidly revolving sling. A moving body will move in a straight line unless some force obliges it to alter its course. Thus it appeared likely that in reality the moon was always falling towards the earth, and that it was this constant falling which prevented it from moving off in a straight line. Newton then proceeded to prove by most ingenious mathematical reasoning that the force of gravity, if it were such as he supposed it to be, would keep the moon constantly moving round the earth. He also showed that, if his hypothesis of gravity were

true, the planets would move round the sun as they do. He went on to explain a great many peculiarities in the motions of the planets and their satellites. He showed that even the comets though they come and go in so apparently irregular a manner, really move in long orbits as gravity would make them move. The tides, too, are another peculiar effect of the same force. Thus his law became a verified hypothesis, one so entirely agreeing with facts that we cannot but believe it to be correct. It becomes an **established law of nature**, and is sometimes called a **theory**, but this last word, theory, is used with several different meanings, and we should take care not to be misled by it. Here it means only a **well-verified hypothesis.**

125. Sometimes it will happen that two or even three quite different hypotheses all seem to agree with certain facts, so that we are puzzled which to select. A little before Newton formed his hypothesis of gravity, the celebrated Descartes had also formed an hypothesis to explain the motions of the heavenly bodies. He suggested that they were carried round in kinds of large whirlpools called vortices, and he pointed out that all the planets go round the sun in the same direction, as they would do in a whirlpool. The satellites of Jupiter, then lately discovered by Galileo, also seemed to go round Jupiter in a small whirlpool, so that the hypothesis was held by many philosophers of the time to be a very good one. Newton's hypothesis of gravity, however, explained the same facts, and it was difficult to decide which was the best hypothesis. That of Descartes was much more simple and easy to understand; that of Newton explained a great many more facts and in a more exact manner.

When there are thus two hypotheses, one as good as the other, we need to discover some fact or thing which will agree with one hypothesis and not with the other, because this immediately enables us to decide that the former hypothesis is true and the latter false. Newton pointed out that comets do not agree in their movements with Descartes' whirlpools, because they pass right through the sun's great whirlpool without moving like the planets which rest in it. Even when a comet passed through the supposed smaller whirlpool of Jupiter, it moved on as if there were no such whirlpool. We now know, too, that great numbers of comets pass round the sun in all directions. Each would require its own separate whirlpool according to Descartes' hypothesis, but as there can be only one great whirlpool round the sun, namely, that which carries the planets, it becomes quite impossible to explain the motions of the comets by Descartes' vortices. All the comets on the other hand, as far as they have been observed, agree with Newton's hypothesis of gravity.

126. When any fact, like the motion of comets in the above case, enables us to select one hypothesis and reject other ones, the fact is called a **Crucial Instance**, because it serves like a Crux or Fingerpost, to point out the road which we should take. When we try an experiment which will decide in favour of one hypothesis and against another, it is called an **Experimentum Crucis**.

XVI.—INDUCTIVE REASONING IN ORDINARY LIFE.

127. It is not only in scientific matters that we use hypotheses in order to learn, by correspondence with facts, what has been happening. We are continually arguing in this way in the commonest affairs, and the mind often goes through all the four steps of preliminary observation, hypothesis, deduction, and verification in a few seconds. For instance, in looking out

of the window into the street of a town, I see that the pavement is wet, instead of being dry as it was an hour before. In all probability I at once consider what can have happened to cause the change. I form several hypotheses: rain may have fallen; a water-cart may have passed down the street; the turncock may have opened the water-pipes in the neighbourhood. With great rapidity I draw deductions from these hypotheses. A water-cart does not usually water the footpaths, but rain would wet the footpath on one side at least. Glancing at the footpaths I see perhaps that they are dry. Rain then is probably not the cause; to be more sure, I glance at the sky, and if I find it apparently clear of clouds, this agrees well with the hypothesis of a water-cart, and I should be finally convinced if I discovered that the wet portions of the street ran in broad parallel lines nearly coinciding with the roadway, or only slightly overlapping the footway, in the manner in which water-carts usually do their work.

128. Inquiries in courts of justice are conducted on exactly the same principles. A burglary has been committed, and the police come to examine the premises. This is preliminary observation. They find that the entrance has been skilfully effected, and at once form hypotheses as to the men supposed to be burglars who are at large. They further inquire as to the appearance of men seen going about in the neighbourhood on the night in question. If any suspected character agrees in appearance with a man seen, he is probably apprehended, because the hypothesis of his guilt has received some slight confirmation. His house being searched is found to contain a jemmy and a few other tools which are used in housebreaking. Surely, then, he is a housebreaker; but, if he is the one wanted, the "jemmy" in question will pro-

bably have been used in breaking open the doors, and will have left a mark which should exactly agree in size and character with the tool producing it. Here is deductive reasoning. The tool is carried to the house and compared with any marks which can be found, and if it agrees there is strong verification.

129. The Tichborne trial was probably the longest and most careful inquiry ever held to decide between two hypotheses. One hypothesis was that a certain fat man, now in Dartmoor Prison, is Sir Roger Tichborne; another that he is identical with a butcher called Arthur Orton. Many persons are said still to believe that he is Sir Roger, but in that case they can have no idea what logic or evidence is. Some people believe that because Roger's mother and some of his brother officers and friends recognised the Claimant as Sir Roger, therefore he is so. But many persons also swore that he was not, and some persons swore that he was Arthur Orton. This kind of evidence is very uncertain; for the man was in any case very much changed by age. Where people disagreed so much in opinion, there was but one way of proceeding safely, namely to deduce a great many little circumstances which ought to be true of the Claimant, things he should remember, things he ought to have done, marks which should appear on his body, if he were really Tichborne. We must compare these with the evidence brought forward, and as far as possible we must make a like comparison with the other hypothesis that the Claimant is Arthur Orton. The more slight and apparently unimportant these circumstances are, the better proofs they make, because it is less likely that an impostor would think of them. Thus, when the Claimant wrote to Lady Tichborne from Australia, he addressed her as Mama, whereas Roger had always addressed her in letters as Mother, and it is against all custom and probability for a man as he grows older to

substitute Mama for Mother. He was unacquainted at first with many things which a man could rarely forget, such as the exact name of his own mother, the number of his regiment, the name of the vessel in which he left England. He was entirely ignorant of French, though Roger was brought up in France; yet he knew some Spanish, picked up during a short residence in South America. Roger had been taught Latin at Stonyhurst, but the Claimant did not know the difference between Latin and Greek.

130. On the other hand there were many slight circumstances which agreed with the hypothesis that the Claimant was Orton. He said he had suffered from St. Vitus' dance, which was true of Orton but not of Tichborne. In his will and journal he mentions people known to the Ortons but wholly unknown to the Tichborne family, and moreover displays entire ignorance of his own Tichborne property. The name of the ship in which he says he left England was the *Jessie Miller*, a ship in which it was proved that Orton had sailed. And when the Claimant reached England he went straight to Wapping and inquired after the old butcher who formerly lived there. It is impossible, however, to give in a few words any idea of the force of the evidence taken in the Tichborne trial, because this force arose from the immense number of slight facts and coincidences, each of little importance in itself, but all collectively making the proof as good as certain. A fibre of hemp will bear only a small weight; but if we twist many fibres into each strand, and unite many strands into a rope, we can make a cable as strong as we like. So, we can verify an hypothesis as completely as any one can desire if we can show that it agrees with a great number of diverse facts.

XVII.—OBSERVATION AND EXPERIMENT.

131. There are commonly said to be two ways in which we gain knowledge of the things around us. **The first way is merely to observe what happens without our interference.** We notice the rise and fall of the tides, and if we remember, or set down on paper, the times at which the tide is highest on several days in succession, we shall learn that high tide is about three quarters of an hour later on each day than on the previous day. If we mark the heights of the tides, too, we shall ascertain that they are greatest at the times of new and full moon. In this and a great many other cases, we cannot in any way govern or regulate the things which we notice. The motions of the stars and planets, the changes of the weather, storms, earthquakes, volcanoes, meteors, are things which go on quite beyond our control. In inquiring about such things, then, we can only employ **simple observation.**

132. When we can manage it, we should **make experiments,** that is, we should put together the things of which we wish to learn the nature, in such a way as to show what the action will be under certain known circumstances. **In experimenting we interfere with things, and then observe the result; experimentation is observation with something more, namely regulation of the things whose behaviour is to be observed.** The advantages of experiment over mere observation are of two kinds.

133. In the first place, we shall generally know much more certainly and accurately with what we are dealing, when we make experiments than when we simply observe natural events. A chemist may very

properly wish to learn the action of carbonic oxide gas upon animals and men, when taken into the lungs. If he trusted to mere observation, he would have to wait until some animal went by accident into a room, well, or other place full of the gas. This would only rarely happen, and when it did happen, we could hardly be sure whether the gas was really carbonic oxide gas; for it would probably be mixed with much carbonic acid gas, which is said to be quite different in its action on living beings. By experiment we should learn all that we want very quickly, because we might fill a glass vessel full of the pure carbonic oxide gas, and put a small animal such as a rat into it, and observe the effects exactly. When so many rats and other animals are killed every day for less necessary purposes, there can be no harm in a chemist killing one or two rats, when he may thereby learn something exceedingly useful to men and animals for ever after. Carbonic oxide gas might be very valuable for warming and lighting houses at small cost, and thus saving the lives of many persons, if it were not apt to do harm by escaping and poisoning people. We do not know how great the risk is, but proper experiments would soon show this.

134. Nature sometimes seems to make experiments for us. Near Naples there is a very curious cave, called the Grotto del Cane. Men can walk safely into it, but dogs when they enter soon fall down and die, unless quickly removed. At first sight, it might appear as if there were some substance in the cave poisonous to dogs, but not to men. A few facts, however, soon negative this hypothesis; for if a man stoop or lie down, so as to bring his mouth within a foot of the floor of the cave, he soon shows signs of suffocation. All that is observed to happen in the cave is easily explained by the fact (Chemistry Primer, Art. 33) that carbonic acid is considerably

heavier than air. A chemist can fill a glass jar with this gas, and then pour it into another jar, almost as he would pour water. A small animal put into such a jar will show signs of suffocation when the carbonic acid is poured in, and this experiment completely explains what is observed in the Grotto del Cane.

135. It is a further advantage of artificial experiments, that they enable us **to discover entirely new substances and to learn their properties.** On the surface of the earth, there is always some chemical action going on among the earth, and sand, and water, but it is the same as has been going on for many thousands of years. It is when we choose particular substances and heat them, or press them, or electrify them in an unusual manner, that we may expect to meet something new. It must have been a surprising discovery when iron was first made from heavy red stones put into a hot charcoal fire; from this and a series of other experiments, we have gained all that iron tools, iron vessels, engines, railways, and steam-boats now do for us. Gold was probably discovered by mere accidental observation, because it is in many places found among the sands of rivers. But mere observation could never have led us to expect that from dull clay we could get a beautiful, strong, and very light metal, named aluminium. It is quite possible that careful and persevering experiments will some day lead to the discovery of an alloy of aluminium, or of some metal now rare or unknown, which will be more useful than all our gold and silver. We must not suppose that we have yet found out the thousandth part of the wonderful things which may be in time discovered by truly scientific reasoning and experiment.

XVIII.—ANTECEDENTS AND CAUSES OF EVENTS.

136. What we want to do, both in observing and experimenting, is **to discover the exact circumstances in which an event will happen.** In other words, we want to know what things must be present in order that something else shall appear. All the objects which are put together in making an experiment, or all the circumstances which precede some natural event, such as a thunder-storm, may be called **antecedents, or the things going before.** All that happens or is produced afterwards are called **consequents.** In the case of the thunder-storm, warm moist air, a bright sun, lofty swelling clouds, and a fall of the barometer, are usually the antecedents, and a heavy shower of rain, lightning, thunder, a squall of cool wind, and a rise of the barometer, are consequents. But it is not to be supposed that all the antecedents of an event will be necessary for its production. The sun might often be shining brightly before a thunder-storm, but sometimes such storms happen in the middle of the night. The sun, therefore, seems not to be needed to produce the storm. If a person be taken ill after eating dinner, all the meats and drinks—beef, potatoes, cabbages, bread, mustard, pepper, salt, water, beer, wine, or whatever else he may have taken—will be antecedents, and his illness is one of the consequents. But it is exceedingly unlikely that there would have been something poisonous in each of the dishes and drinks. What we shall need to do in such a case is to find out in what particular substance the poison was contained, which was the necessary antecedent, or, as it is usually called, the cause of his illness.

137. **The cause of an event is that antecedent, or set of antecedents, from which the**

event always follows. People often make much difficulty about understanding what the cause of an event means, but it really means nothing beyond **the things which must exist before in order that the event shall happen afterwards.** Sometimes it may seem as if one single antecedent is the sufficient cause. If there be copper in the pickles eaten at a meal, it may seem to be the sole cause of the illness of the eater. But the peculiar formation of the stomach, which becomes deranged by the presence of copper, is also a necessary antecedent. Copper does not poison us when we merely go near it. A single spark may seem to be the cause of the explosion of a barrel of gunpowder; but then the gunpowder is equally the cause of explosion, and several substances are requisite to make gunpowder. We shall in vain attempt to produce an explosion with charcoal, or saltpetre, or sulphur taken separately. But if we grind them all up together in particular proportions, and make the mixture into grains, we get something which will explode, that is, very rapidly burn, when a spark falls upon it. Thus the sulphur, the saltpetre, the charcoal, the particular form of the grains, the spark, and, it may be added, the absence of moisture, are all **necessary antecedents or causes** of the explosion.

138. **The great rule in making experiments is to vary one thing at a time.** Our purpose is to ascertain exactly which antecedents of an event are requisite to produce it. But if I alter two or more antecedents at the same time, and the result is altered, I cannot tell whether the change is due to one antecedent, or to the other, or, it may be, to both. If a cup of tea does not taste well, it may be due either to the poor quality of the tea, or to the water not boiling when it was made. If I have a new pot of tea made with boiling water and a different kind of tea, I may get a better cup of tea, but I shall not learn why the former

cup was bad. I must first try the original kind of tea with boiling water, and if it still tastes bad, I shall know that the fault was in the tea.

If a person in perfect health falls down stairs, and receives severe injuries, followed by death, we feel sure that the fall caused the death. But if a person is seized with some kind of fit and then falls down, and dies soon after, the fatal result may be due either to the fall or the fit, or to both, and the minutest inquiry may hardly settle which is the case.

139. Every one knows that a bright piece of iron soon rusts when exposed to the air. What are the causes of this rusting? If we put a piece of bright iron into a glass tube, exhaust the air out of it, and seal the tube up, the brightness of the metal will remain undimmed for any length of time. But air is a mixture of oxygen, nitrogen, vapour of water, carbonic acid, and small quantities of other substances. The air always contains, too, a very slight quantity of common salt, in small particles which float about. Any of these substances, then, may be causes of the rusting of iron, and to decide which are the causes, it is not sufficient to withdraw air altogether, nor even to try pieces of iron with pure oxygen, nitrogen, and vapour of water separately. It will be found that the iron does not rust with any of these substances when quite pure. The most instructive experiment is to take common air and remove all the moisture from it; iron will remain perfectly bright in such air, so that **moisture is one of the causes of rusting.** But it is not the only cause; for in perfectly pure water, or vapour of water, free from oxygen and carbonic acid, iron also remains bright. In a mixture of oxygen, watery vapour, and carbonic acid, such as air would be without the nitrogen, iron rapidly rusts. By further similar experiments we should be led to conclude that two substances, oxygen and vapour of water, are

necessary antecedents of the rusting of iron, and that carbonic acid, if not altogether necessary, makes iron rust much more rapidly. This instance shows that it is not always easy to find out exactly which of the many antecedents of an effect are the necessary antecedents or causes of the effect.

XIX.—DISCOVERY OF AGREEMENT.

140. What we want to do both in observing and experimenting, as we have learnt in the last Article, is to discover the circumstances which always precede an event. The first step towards this discovery is usually to try and find out what there is alike in the antecedents of every particular case when the event occurred. Accordingly, when we wish to explain the occurrence of anything, **we should begin by thinking of everything like it that we have ever seen or heard of**, and then we should compare these things together carefully, and try to detect the exact likenesses between them.

141. Suppose that we see a bright rainbow in the sky, and want to learn exactly why it occurs then and not at other times. We want to know, in short, what are the causes of its occurrence. We must begin by comparing together all the occasions we can remember when a rainbow was seen. We may observe that whenever such a bow appeared, rain was falling somewhere in the sky. As the name implies, the rainbow always occurs on or among rain drops, and no one ever saw a rainbow with a perfectly clear sky. At the same time, clouds and rain must not obscure the whole sky. The sun must be shining while the rain is falling. We may easily remember that rainbows occur with occasional brief showers of rain, or when a storm is nearly at an end, and the sun is beginning to shine forth again.

142. We ought not to content ourselves with considering ordinary rainbows only; we should think and collect information about all cases in which similar coloured bows, or even similar colours, are produced. Lunar rainbows are sometimes seen, and when seen there is a bright full moon shining on a shower of rain. Comparing lunar with solar rainbows, we find that the sun is not requisite, but that any bright beam of light shining upon a shower of rain seems to be the necessary antecedent. Nor is rain falling from the sky quite necessary. Some waterfalls—especially the Rjukan or Smoking Foss in Norway—throw up clouds of fine spray composed of minute particles of water. If we see the sun shining in a particular direction upon such spray a bright bow, exactly like a rainbow, is discovered. The fine drops of water from a fountain occasionally show fragments of a similar bow. In the early morning the grass, and shrubs, and spiders' webs are sometimes covered with drops of dew, and a bright sunbeam produces upon them a rainbow turned upside down. At sea the colours of the rainbow may be seen upon the spray as it is driven above the surface of the sea by the wind after a storm.

Comparing the different occasions on which the same sort of bow is seen, we discover that **a beam of light and particles of water, in a particular position are the necessary antecedents or causes of the bow of colours.** This is nearly all that simple observation can tell us, and it forms merely the first step of preliminary observation.

143. It was Sir Isaac Newton who fully explained how rainbows are produced, and this he did by means of hypotheses. Long before his time, indeed, it was remarked that colours, similar in their succession to the seven colours of the rainbow, are seen in sharply cut glass vessels, diamonds, or other transparent objects. Roger Bacon, whom I mentioned before

(Art. 110), had discovered the circumstances in which a rainbow appears, and had also remarked the resemblance to the colours of crystals. Another early experimenter pointed out that similar effects are produced by a sunbeam falling on a glass globe full of water. But Newton did a great deal more; for he imagined the different ways in which a ray of light might enter a drop of water and get out again, so as to reach the observer's eye, after having been reflected and refracted within the drop. Knowing the laws of the reflection and refraction of light, he was able to calculate the angle between the ray coming out and that going in, and thus to decide the size and position of a rainbow, with respect to the sun and the eye of the observer.

144. Measurements of rainbows agreed with Newton's calculations; but he was not contented with this verification alone. He proved that a second, but smaller portion of the light entering a drop of rain, would come out in a different direction, so as, when bright enough, to form another larger rainbow. It is well known that a rainbow when very brilliant is often accompanied by a second fainter bow, and in this we have a complete verification of Newton's theory. In such a case we can see clearly how philosophers, beginning with simple preliminary observation, gradually went through all the steps mentioned in Article 118, and, by hypothesis, deduction, and verification, arrived at a true theory.

XX.—THINGS WHICH VARY IN QUANTITY.

145. The causes and effects with which we have to deal in science can often be made to vary in quantity. We can make a body more or less hot or cold; we can put a greater or less weight to press upon it; or we can try how much a magnet of greater or less force will

attract it. Whenever we can thus alter the quantity of the things experimented on, we can apply a rule for discovering which are causes and which are effects. **We must vary the quantity of one thing, making it at one time greater and at another time less, and if we observe any other thing which varies just at the same times, it will in all probability be an effect.**

We may easily observe, for instance, that when air is forced into a fire by use of the bellows, greater heat is produced; the more powerfully we blow, the hotter the fire becomes, and as soon as we leave off blowing, the fire begins to cool. There can be no doubt, then, that a supply of air is one of the causes of the combustion of fuel. In the same way we may easily prove that sunlight is one necessary condition of the growth of plants. The sun partly makes the experiment for us in this case, because it shines so much more powerfully, and for a longer time, in summer, than in winter, and we see that grass and plants grow rapidly in June and July, and hardly at all in December and January. But this is not quite satisfactory, because the air is much warmer in summer than in winter, and this might perhaps be the reason.

To satisfy ourselves, we ought to make more exact experiments, by taking several plants of exactly the same kind, planted in similar pots containing similar soil, putting some plants where they will receive bright sunshine, others where they will be partially shaded, as under trees, and some again under boxes or in sheds, where they will have little or no light, but where the air will be of the same temperature as outside. Then, as nearly as can be expected, the growth of the plants will be found to correspond to the quantity of sunshine falling upon them.

146. From the foregoing example we may learn the need of the precaution, **to vary only one thing at**

a time, as far as we can possibly manage it. This is, in fact, the same precaution which we had to take in simple experiments (Art. 138), putting one thing in operation at once. Now we must make one cause greater and less, keeping all the other things which are present of the same quantity as exactly as we can. If we were to put one plant where it would have both more sunshine and more moisture than another similar plant, we could not be sure whether the difference of growth was due to the difference of sunshine, or to the difference of moisture. If possible then we should try plants having equal quantities of moisture, and in every other respect alike, with different quantities of sunshine. Then again, if we want to know the effect of moisture, we should take similar plants, similarly supplied with sunshine, and differing only in the supply of moisture.

XXI.—THINGS WHICH VARY PERIODICALLY.

147. The changes and motions which things about us exhibit are often what we call **periodic**, that is, they happen over and over again in a similar manner after equal periods or intervals of time. Day and night are periodic changes, for they happen alternately, and one night is nearly equal in length to the preceding or following night. But, as summer approaches, the daylight grows longer, and the nights shorter; this happens in almost exactly the same way every year, so that it is also a periodic change, depending upon the motion of the earth round the sun. The tides also rising twice a day are periodic.

148. When things thus vary regularly and frequently, there is a simple rule, by following which we can judge whether changes are connected together as causes and events. **Those things which change in exactly equal times are in all likelihood**

connected together. Almost every day the air becomes warmer in some degree during the afternoon, and when we take the average of several weeks or months, we find that it is almost always warmest about three o'clock in the afternoon. There can be no reasonable doubt, of course, that this increase of heat is caused by the sun, which is at its highest point in the heavens about twelve o'clock, but continues to warm the air more than it is cooled, for three hours afterwards. In the same way the warmest day in the year is about the 21st July, and this is, on the average, at an equal interval from the 21st June, the longest day. Even if we did not on other grounds know it to be the case, we should infer that the warmth of summer is due to that periodic motion of the earth round the sun, which causes the sun to shine longer and brighter during summer than during winter.

149. In other cases we learn from periodic changes that most unexpected things are connected together. I have mentioned the tides as periodic events; now, as the tides happen at intervals of about $12\frac{3}{8}$ hours, whereas the sun goes round the heavens at intervals of about 24 hours, we cannot conclude by our rule that the sun is the cause of the tides in question. We have to look out for some other cause which varies, or moves round in $12\frac{3}{8}$ hours. We should not meet with anything exactly answering this description; but we should find that the moon gets nearly to the same place in the heavens on successive evenings at intervals double that named, or $24\frac{3}{4}$ hours.

When the moon is quite new, it is seen early in the afternoon; but as it grows older and older, it rises later, until at last it is not seen at all till early morning. If, when conveniently seen in the evening, we noted the time of its reaching a certain position in the heavens we should find the time to be three quarters of an hour later every night. The tides are just so much later also; hence it becomes very probable that the attraction of the moon on the ocean is the cause of the tides. Sir Isaac Newton showed beyond all doubt that this was the case, and he explained why there are two tides in the $24\frac{3}{4}$ hours instead of one tide.

150. In the last thirty or forty years very curious discoveries have been made about variations in the atmospheres of the sun and earth. It was well known to Sir William Herschel and other astronomers, seventy years ago, that the spots on the sun's face are much more numerous and large in some years than in others. Careful observers having registered the spots for many years, discovered by degrees that the years in which the spots are very numerous, happen at intervals of about eleven years. There were a great many spots in 1837, in 1848, in 1859, and in 1870, and comparatively few in the intermediate years, about 1842, 1853, and 1864. It was also noticed that those wonderful and unaccountable displays of light in the heavens, called Auroras, are much more frequent and grand in some years than in other years. Strange to say, when there are many sun-spots there are many fine Auroras, as in the autumn of 1859, and again in 1870. It is impossible to say at the present time how spots in the sun can cause Auroras; but they vary together so regularly that there can hardly be any doubt about their being connected together.

There is now reason to believe that the typhoons, or great storms which occur in parts of the tropical regions of the earth, also depend upon the sun-spots. Meteorologists are endeavouring to discover whether the comparative coldness or warmth of some years, or the variations in the quantity of rain, may not also have some connection with the spots on the sun, but we ought to be very careful in drawing conclusions about such uncertain changes. Sir William Herschel thought that the variations in the price of corn depended upon those of the sun-spots, and this, if proved, would be a very interesting and important discovery. I have tried to ascertain whether it is so or not, but have been unable to find any evidence of the truth of Sir W. Herschel's hypothesis.

XXII.—REASONING FROM EXPERIMENTS.

151. It would be a mistake to suppose that the making of an experiment is inductive reasoning, and gives us without further trouble the laws of nature. **Experiments only give us the facts upon which we may afterwards reason.** If I wrap up a piece of ice in a blanket and, placing it alongside of another piece of ice not wrapped up, observe that the latter rapidly melts away, and the former does not, there are only two observations here. If I were to draw the conclusion that a piece of ice wrapped up in a blanket always melts less rapidly than one not wrapped up, this would be a case of inductive reasoning, but a bad case, because it would not always hold true. If the temperature of the surrounding air, and of other objects, were below the freezing point, neither of the pieces of ice would melt.

152. Experiments then merely give facts, and it is only by careful reasoning that we can learn when the same facts will be observed again. **The general rule is that the same causes will produce the same effects.** Whatever happens in one case will happen in all like cases, provided that they are really like, and not merely apparently so. The advantage of being able to try experiments is that we ascertain exactly what are the antecedents and surrounding circumstances of an experiment, and we can vary these so as to find out which are important and which are not. If we wished to decide exactly in what circum-

stances the melting of ice would again be observed, we should have to mark the temperature of the air, and try the experiment over and over again at different temperatures. We should also have to consider whether the sun was shining, or whether heat could reach the ice from fires, or warm bodies in the neighbourhood.

153. When we have by repeated experiments tried the effect which all the surrounding things might have on the result, we can then reason with much confidence as to similar results in similar circumstances. But we can never be quite sure about the matter. It is always possible that we have overlooked the thing which is really necessary to the result of the experiment. It may be very unlikely, but it is possible. Now and then chemists find that some experiment which they thought they understood completely, deceives them and gives quite unexpected results. Sometimes they can afterwards explain these exceptions and failures. They may happen to have met with a new substance which looked like another substance familiar to them, but was really different in its properties. This is the usual way in which new elements are discovered.

154. In order that we may, from our observations and experiments, learn the laws of nature and become able to foresee the future, we must perform **the process of generalization. To generalize is to draw a general law from particular cases, and to infer that what we see to be true of a few things is true of the whole genus or class to which these things belong.** It requires much judgment and skill to generalize correctly, because everything depends upon the number and character of the instances about which we reason.

XXIII.—HOW AND WHEN TO GENERALIZE.

155. It is very difficult to explain how it is that we can ever reason from one thing to a class of things by generalization, when we cannot really be sure that the things resemble each other in the important points. A wine merchant generalizes on a small scale, when he takes a single glass out of a pipe of wine, and infers that the quality of every other glassful drawn from the same pipe will resemble this particular glassful. But then he knows that the wine in the pipe has been well mixed up, so as to be exactly alike in all parts. Similarly a broker who sells cotton, corn or sugar, has a sample taken which fairly corresponds to the whole of the lot of goods, and the buyer takes the goods on the belief that the sample is really a fair one.

156. Who is to say what is a fair sample of things in nature? Can we say that, because all the stones observed by us fall to the ground again when thrown up, therefore all other stones will do the same? If so, upon what grounds do we argue? We have to get a general law from particular facts. In reality this can only be done by going through all the steps of inductive reasoning as explained in Articles 112 to 118. Having made certain observations, we must frame hypotheses as to the circumstances, or laws from which they proceed. Then we must reason deductively, and, after verifying the deductions in as many cases as possible, we shall know how far we can trust similar deductions concerning future events. But this long process has been performed very frequently by philosophers, and it usually leads to the conclusion, that **things which resemble each other in several of their properties will probably resemble each other in more properties.** **There is no certainty in the matter,**

and as I have already said, it is difficult to judge when we may, and when we may not, safely infer from some things to others in this simple way, without making a complete theory of the matter.

157. The only rule that can be given to assist us is that **if things resemble each other in a few properties only, we must observe many instances before inferring that these properties will always be joined together in other cases.** We notice that stones when thrown into the air, fall to the ground, and the same is true of pieces of wood, metal, ice, leaves of trees, feathers, or scraps of paper; even spiders' webs, and the lightest things do the same when not prevented by wind. All these are material solid bodies, and we may observe that the circumstance of falling to the earth does not seem to be connected with the colour, size, shape or other peculiarities of the things. The things, in short, which fall, resemble each other in no apparent circumstances except that they do fall, and that they are solid and material. Further observations show that liquids also fall, as in the case of rain. Clouds, smoke, steam, and dust seem not to fall, but further inquiry shows that in all these cases the particles are really falling as rapidly as the air will allow them. Moreover, the air itself falls very rapidly, when there is an empty space or vacuum into which it can fall. Thus we find that even solidity is not necessary to the property of falling, but that all bodies, which consist of matter at all, also have weight. These circumstances having been so often joined together, we are justified in expecting that they will be joined together in all future cases which we may be able to observe. We conclude, therefore, that all material bodies will have the property of falling in the same manner as the stones and other things observed. In other words, we learn the general law that all things which resemble each other in being material,

will also resemble each other in the property of falling towards the earth, when not prevented by any other force. This is a very perfect instance of generalization, and the conclusion has been confirmed by Newton's hypothesis of gravitation, and the observations made on the motions of the heavenly bodies.

158. As a second instance of a good generalization, let us consider what we can infer about the bright colours seen upon soap bubbles. If we were to generalize carelessly, we should perhaps infer that all soapy water ought to show bright colours; but on examining the soapy water which we used, we should find ourselves wrong. To know when to expect similar colours, we must take every opportunity of observing the same thing again. When tar is spread in a thin film over water, as may sometimes be seen in canals and docks, it also shows most beautiful colours of the same kind. Now the film of tar does not seem to resemble a soap bubble in anything but being very thin. When a piece of thick glass is cracked, and we examine the crack very carefully, we shall often find colours similar in appearance, though perhaps less brilliant; and, if we press two plates of glass together, or still better, press a nearly flat lense upon a piece of plate glass, colours are seen near the place where the two pieces of glass touch. It is difficult to say in what way tar, soapy water, and cracks in glass resemble each other, unless it occurs to us that between the two surfaces of the glass there is a thin space filled with air. The colours thus appear in three cases where light falls upon a very thin film of substance with two bright surfaces close together. Further inquiry would show that this was a good case for generalization, and that any very thin transparent plate upon which light falls, will produce similar colours. When we see such colours, then, we may expect that

there will be found thin plates of substance. The bright colours of mother-of-pearl arise in this way from the extreme thinness of the layers of which the shell is formed.

XXIV.—REASONING BY ANALOGY.

159. At the beginning of this Primer, I described the way in which we commonly reason, from one thing directly to another (Articles 4 to 6), as from the mountains of California to those of New South Wales, or from one orange to another. This kind of reasoning may be called **Reasoning by Analogy**, and it only differs in degree from that kind of reasoning called generalization. **When many things resemble each other in a few properties, we argue about them by generalization. When a few things resemble each other in many properties, it is a case of analogy.** If only a very few things resemble each other in a few points, we should have no ground for arguing from them to other things. But when there are either a number of things showing resemblance, or a number of properties in which they show resemblance, then we have some grounds for inferring that the same properties will be found joined together in other cases. **The rule for reasoning by analogy is, then, that if two or more things resemble each other in many points, they will probably resemble each other also in more points.**

160. If I see a machine with boiler, cylinder, air-pump, piston-rod, crank, and other parts exactly resembling those of a steam-engine, I do not hesitate to call it a steam-engine, to assert that it has a piston, valves, and other hidden parts, like all steam-engines. It is in the same way that we reason about the sub-

stance of which anything is made. If a person offers me a shilling as change, how can I be sure that it is a good shilling, and made of silver? All that I can do is to examine the coin, and observe whether it has a fine pure white lustre where the surface is rubbed; whether there is in other parts of the surface the black tarnish peculiar to silver; whether the coin seems to be hard, and gives a sharp ringing sound when thrown down. If it has all these characters and, moreover, has a good impression exactly like that seen on other shillings issued from the mint, then it is doubtless made of silver, and is a true shilling, that is to say, it will show all the other properties of standard silver, when examined in a manner suited for showing them.

161. In spite of the very distinct marks by which we may usually recognise a silver coin, we know that counterfeit ones are often made and passed from one person to another. In these and many other cases **reasoning by analogy is found to be a very uncertain guide.** In some cases unfortunate mistakes are committed. Children are sometimes killed by gathering and eating poisonous berries, wrongly inferring that they can be eaten, because other berries, of a somewhat similar appearance, have been found agreeable and harmless. Poisonous toadstools are occasionally mistaken for mushrooms, especially by people not accustomed to gather them. In Norway mushrooms are seldom seen, and are not eaten; but when I once found a few there, and had them cooked at an inn, I was amused by the people of the inn, who went and collected toadstools and wanted me to eat them also. This was clearly a case of mistaken reasoning by analogy. Even brute animals reason in the same way in some degree. The beaten dog fears every stick, and there are few dogs which will not run away when you pretend to pick up a stone, even if there be no stone to pick up.

162. In science a great deal is learnt by analogy. We know that the moon has mountains, because there are marks on the face of the moon, which closely resemble the appearances which our mountains would have as seen from the moon. The moon's mountains cast longer shadows as the sun is setting, and shorter ones as it is rising, just as it happens on the earth's surface. But the ancient astronomers were misled by analogy into thinking that the flat dark spaces on the moon's surface were seas; they thought that the moon would naturally have oceans and seas of various sizes, like the earth. By the use of large telescopes we now know that there are no seas, rivers, or other preceptible bodies of water on the moon. (Primer of Astronomy, Art. 129.)

163. Sometimes the analogy between things is so complete and exact that we cannot doubt it for a moment. The Chinese have printed mathematical tables of numbers called logarithms; but on examining these tables they were found to have the same mistakes as some English tables of logarithms. The analogy was so complete that we must believe the Chinese tables to be copied from the English ones. This is the only hypothesis which can explain the resemblance. As we walk over the flags in a street, we may often notice that the surface is wavy, in a manner exactly resembling a fine sandy sea beach, from which the tide has just receded. Sometimes we may notice on flagstones little pits or hollows, alike in form and size to the holes which large drops of rain make in a sandy surface. The tracks of insects also and the foot-prints of birds, and other animals are sometimes seen. We cannot explain these precise analogies between the flagstones and the sea beach, except by supposing that the flagstones really were formed of the sand and mud deposited by waves upon a sea beach countless ages ago. Geologists continually argue by analogy in this

way from what goes on under their eyes in the present day to what must have happened when the hardest rocks were being slowly formed.

164. Of all the planets Mars seems to have the closest analogy to the earth. When carefully examined it is found to have darker portions, believed to be seas, and lighter portions which are probably land. At each pole of the planet, too, is a white round patch; now each of these patches, if carefully watched, is found to decrease when Mars is in such a position as to expose the spot to the sun's rays, and to increase at other times. These white spots thus behave exactly like the masses of snow and ice at the north and south poles of the earth. The analogy is so perfect that we conclude, almost beyond doubt, that Mars has regions of ice and snow at its poles like the earth. (Primer of Astronomy, Art. 162.)

165. **There is no way in which we can really assure ourselves that we are arguing safely by analogy.** The only rule that can be given in this, that the more closely two things resemble each other, the more likely it is that they are the same in other respects, especially in points closely connected with those observed. Not only is it very probable that the spots on Mars are composed of ice and snow, but we may also infer that Mars has an atmosphere with winds, clouds, rain, and other things very like our own. Some people argue, too, by analogy that there are probably living beings on Mars more or less resembling the plants and animals on the earth; but it is evident that reasoning on such a matter is very uncertain. In order to be clear about our conclusions, we ought in fact never to rest satisfied with mere analogy, but ought to try to discover the general laws governing the case.

166. In analogy we seem to reason from one fact to another fact without troubling ourselves either with

deduction or induction. But it is only by a kind of guess that we do so; it is not really conclusive reasoning. We ought properly to ascertain what general laws of nature are shown to exist by the facts observed, and then infer what will happen according to these laws. This we can do in the case of the white spots on Mars to a great extent. We know very well that the rays of the sun melt snow and ice, and we observe exactly how in the Arctic regions these effects take place. We are therefore prepared to explain the increase and decrease of the white spots of Mars by reasoning deductively. But this does not apply to the supposed inhabitants of Mars. No one has ever been able to discover how living beings came to exist on the earth, and no one can be proved to have produced a living creature out of dead matter. We cannot therefore argue deductively that living beings would be produced on Mars, because its surface and atmosphere are in some ways like those of the earth.

167. In other matters people are continually led into errors by trusting to slight analogies. A few years ago it was common to hear it asserted that the government would make profit by sending telegrams at very small charges. It has even been said that the railway companies ought to carry passengers any distance at the same low charges which are required for letters and books. These people point to the Post Office as an institution which earns for the government a large profit although it only charges a penny for a letter, and a halfpenny for a card or newspaper. They say, too, that as the prices of the daily newspapers were in past years gradually reduced from sixpence to one penny, the proprietors got larger profits. Then by analogy they infer that the same will happen with telegraphs and railways. But this is a mere guess, and a very bad one. They ought not to be satisfied with mere apparent resemblance, but should inquire

into the reasons why the penny post and the penny newspapers pay so well.

168. They would find, for instance, that it is not the pennies paid for newspapers which make the profits of the publishers, but the large sums of money which are received for advertisements. In telegraphs and railways there is little or no source of profit analogous to advertisements. They would find, again, that the Post Office is very profitable to the government, because a postman can carry a great many letters and cards at the same time, and can deliver a bundle of half-a-dozen almost as quickly as a single one. The Post Office, therefore, can usually do more work without employing more men, and the more letters it delivers the greater is the profit. With the telegraphs, however, it is quite different. A clerk cannot telegraph a dozen messages along the wires at once, nor even two messages. Each message has to be sent separately and delivered generally by a messenger employed for this single purpose. The more messages are sent the more clerks and messengers are needed. If the charges were to be made very low, the government would lose a great deal, instead of gaining as they do in the Post Office. We find then that reasoning by analogy is not to be depended upon, unless we make such an inquiry into the causes and laws of the things in question, that we really employ inductive and deductive reasoning.

XXV.—FALLACIES.

169. In learning how to do right it is always desirable to be informed as to the ways in which we are likely to go wrong. In describing to a man the road which he should follow, we ought to tell him not only the turnings which he is to take, but also the turnings

which he is to avoid. Similarly it is a useful part of logic which teaches us the ways and turnings by which people most commonly go astray in reasoning.

170. **Errors and mistakes in reasoning are called fallacies,** that is, modes of reasoning which deceive. But we ought not to confuse a false opinion with the bad reasoning by which it is reached. The word fallacy is in fact an ambiguous one (Art. 29). In one sense it is a fallacy that the moon governs the weather, because long and careful inquiries have shown that there is no correspondence between the changes of the moon and the changes of the weather. But this is a fallacious or false opinion: the logical fallacy consists in the bad reasoning which has by degrees led people to believe in the moon's power. On one or two occasions a person may notice a change of weather on the day of new moon, and he thinks it so singular that he tells his neighbours of the fact, and they remember perhaps to have noticed the same thing once or twice. But it is bad reasoning to argue that, because on a few occasions things happen one after the other, therefore the one is the cause of the other.

171. There are at least twelve new moons in each year, and changes of the weather take place in this country at least once a week on the average. It is therefore quite likely that a new moon and a change of the weather will happen together now and then. But most people believe that the moon affects the weather not because they have really noticed it to be so, but because they have often heard it said to be so. This is not bad reasoning, like that which gave rise to the false belief, but it is simply repeating the same false opinion. In logic we ought to use the word fallacy to mean only false reasoning, and not false beliefs.

172. Taking the word fallacy, then, to mean bad reasoning, we must remember that several different ways of falling into erroneous reasoning were described

in the Articles on deductive logic. Whenever we break the rules for converting propositions, the rules of the syllogism, or any of the other rules which were given for guiding us in making inferences, we commit a fallacy. If we infer that, because all the ordinary animals known to us have the power of moving themselves, therefore this object which has the power of moving itself is an animal, this is against the third rule of the syllogism, and is a case of the fallacy of undistributed middle term (Art. 85). Each of the other rules of the syllogism, when broken, gives rise to a distinct kind of fallacy: a breach of the first rule is called a Fallacy of Four Terms: if we attempt to draw a conclusion from two negative premises, there is said to be a Fallacy of Negative Premises. In these and some other cases the badness of the reasoning ought to be apparent to any one who has carefully studied what I have said about the syllogism. But an argument may seem to agree with the rules given and yet may be fallacious, owing to some confusion in the meaning of the terms or propositions. We must consider in what ways such fallacies are most likely to arise.

XXVI.—FALLACIES OF AMBIGUITY.

173. Perhaps the most common cause of bad reasoning is the **use of ambiguous terms**, which mean one thing in one place and another thing elsewhere. **A word with two distinct meanings is really two words.** If a person were to argue that his ailment is a cold, and that all cold is dispelled by heat, therefore his cold will be dispelled by heat, it would be absurd thus to confuse together a cold or catarrh with the absence of heat. To argue thus is as bad as having four terms in the same syllogism, and comes

in fact to the same thing. But in many cases it is by no means easy to see that we are using the same word with two meanings.

174. It has recently been argued that since all mendicants can be punished by law, and Sisters of Charity who ask for subscriptions are mendicants, therefore Sisters of Charity who ask for subscriptions can be punished. On the same grounds, however, anyone who goes about soliciting subscriptions for a charitable purpose, would be liable to be sent to gaol as a rogue and vagabond. A mendicant is no doubt one who begs; but we must not convert this proposition simply, and say that whoever begs is a mendicant. A true mendicant not only begs, but lives upon what he gets by begging, and does no useful work in return. When, therefore, the law punishes mendicancy, we must take care that it is applied only to those who beg for their own support, and make themselves a nuisance to the public. Lawsuits frequently arise from the difficulty of deciding exactly what words mean. A kind of dull black shaly rock has in late years become very valuable because it can be used to make petroleum. Some of this mineral, known as the Boghead coal, having been found in an estate in Scotland, a great lawsuit took place to decide whether it was or was not really coal. The uncertain meaning of a word may sometimes be the cause of war between great nations. The long dispute between the United States and England, about what was called the Alabama Case, turned on the meaning of the expression "to equip a ship of war." International law allowed the building and selling of ships of war, provided that they were not sent out fully equipped for fighting; but there were differences of opinion as to what equipping meant.

175. At the time of the French Revolution some philosophers argued that kings and rulers ought to do exactly what the people like, because they are the

"servants of the people," and servants should obey their masters. But here is an obvious fallacy of ambiguity. Kings and rulers ought, no doubt, to serve their people, in the sense of doing what is on the whole most beneficial to the people. But there is little or no analogy between service in this sense, and the service of footmen, porters, and domestic servants generally who are paid to give aid to their employers when desired. People fall into a somewhat similar confusion of ideas when they think that, because a member of parliament is elected to represent a certain borough or county, therefore he is bound to vote according to the wish of the people who elected him.

176. There are, indeed, several kinds of fallacy arising from ambiguity, which may be more or less exactly distinguished. Sometimes the **confusion arises between a term in its collective and its general meaning,** and I pointed out in Art. 17, the need of bearing in mind the existence of collective terms. It would be obviously absurd to argue that because all the books in the British Museum Library are sure to give information about King Alfred, therefore any particular book will be sure to give it. By "all the books in the British Museum Library," we mean all taken together. There are many other cases where the confusion is not so evident, and where great numbers of people are unable to see the exact difference. The absurd clamour about the Tichborne trial probably arose from people thinking that, because almost any witness brought against the claimant may be mistaken, therefore the whole of the witnesses taken together may be mistaken. Looking, again, to the things said and done by the claimant, it can be urged, that he may have forgotten the French language; he may have forgotten the name of his mother; he may have mistaken the number of his regiment; he may have confused the name of his ship with that of another

ship; and so on, through the hundreds of facts brought out at the trial. But though a man, under the circumstances, might have done any of these things, **it is exceedingly unlikely, and indeed quite inconceivable, that he should have done all of them together,** had he been really Sir Roger Tichborne. It is the collecting together of a great many slight and independent facts, which sometimes makes circumstantial evidence, as it is called, as complete a proof as can be needed.

177. It may be shown that members of trades-unions often fall into a fallacy of the same kind. They argue that stone-masons, by limiting the number of apprentices, may raise their own wages; carpenters can do the like; and also brickmakers, engineers, cotton-spinners, and so on through the whole list of trades. It is quite true that any one trade may do so to a certain extent; but it does not follow that all trades taken together can do it, because each trade, in thus raising its own wages, tends to injure the others in some degree. We may see in this and many other cases, that a logical distinction, which seemed absurdly obvious when first stated, may really be overlooked by immense numbers of men, and the confusion gives rise to very great harm.

178. It is probably a fallacy of this kind, too, which leads persons to argue that a very rich man ought to give a handsome subscription to a particular institution, because he would never feel the loss. It may be quite true that he would never feel the one subscription solicited, but exactly the same argument might be used in many other cases. The richest person would soon be ruined by the great number of demands which could be made on the same grounds. What a subscriber must look to is not the effect of each separate subscription, but of the whole of the subscriptions which may be expected from him.

179. We sometimes fall into the opposite fallacy to that last described, and argue that, because something is true of the whole of a group of things, therefore it is true of any of those things. It is the **fallacy of arguing from the collective to the general.** All the soldiers in a regiment may be able to capture a town, but it is absurd to suppose that therefore every soldier in the regiment could capture the town single-handed. White sheep eat a great deal more than black sheep; but that is because there are so many more of them. Ministers sitting in Cabinet Council will probably come to a wise decision concerning an important question; but it does not follow that any one of them alone would come to a wise decision.

180. Moral teachers are fond of encouraging us with various good proverbs, such as "Labor omnia vincit." It is difficult to say exactly what is meant by "Labour overcomes all things," unless it be that a sufficient amount of labour will accomplish any practicable scheme. But of course it does not follow that, because a great collective amount of labour will build a pyramid, or make a canal, or compile a cyclopædia, therefore a single person's individual labour can do such tasks. The proverb has little or no value, because every person can give his own meaning to "all things." It is said again, that "what man has done, that man can do." As I am a man I might infer logically from these premises, that I can swim across the Channel like Captain Webb, or write a Paradise Lost like Milton, or discover a new way of making steel like Bessemer, or conquer an empire like Clive. The only way in which the proverb is really true is that, among a collection of a great many millions of men, we can find those who can do all these things. Proverbs often seem very wise, because they are very ambiguous.

181. Other fallacies arise, not from the confusion in

meaning of any one term, but **from the uncertain meaning of a whole sentence.** There is a humorous way of proving that a cat must have three tails: Because any cat has one tail more than no cat, and no cat has two tails, therefore any cat has three tails. As another instance of the way in which we can put nonsense into the form of an apparently good syllogism, take the following: No kind of spirituous liquor ought to be drunk in excess; but water is no kind of spirituous liquor: therefore water ought to be drunk in excess. It seems as if "no kind of spirituous liquor" made a good middle term; but it is not so, and there are really two negative premises from which we can conclude nothing (Art. 81).

182. A common kind of fallacy with orators and those who have to make the best of a bad case, is **proving the wrong conclusion,** and leaving people to imagine, in a confused sort of way, that the case is established. This was the device of the Irishman, who was charged with theft on the evidence of three witnesses, who had seen him do it; he proposed to call thirty witnesses who had not seen him do it. Equally logical was the defence of the man who was called a materialist, and who replied, "I am not a materialist; I am a barber." The officious friend who gives advice is likely to be reminded of the proverb about preaching and practising. But even a drunkard may properly denounce the evils of tippling, and there is no direct connection between the logical strength of an argument and the characters of those who use it.

183. One very dangerous kind of fallacy, not much noticed in books on logic, but of somewhat the same kind as the last named, is **the fallacy of supposing that the failure of an argument tends to prove the opposite conclusion.** Old Mr. Weller, as we all know, had the highest opinion of an "alibi;" but

lawyers say that nothing turns a jury so much against a prisoner as the breakdown of an attempt to prove an alibi. William Sykes being charged with burglary at Bow at one o'clock in the morning, brings witnesses to prove that he was in Whitechapel at that time; but in cross-examination it turns out that, at the best, he is proved to have been at Whitechapel at midnight, so that he might have been at Bow by one o'clock. The jury are apt to assume that therefore he was not at Whitechapel at one o'clock, but at Bow. Yet, unless deduced from something in the character of the witnesses, or the obvious bad faith of the attempt, there is no logical force in the inference whatever.

184. **No number of failures in attempting to prove a proposition really disprove it.** There is a general law of mechanics known under the name of the parallelogram of forces, which is undoubtedly true. A great many ingenious philosophers have puzzled their brains, and written books to prove it true, but none of them have succeeded, except by assuming some other almost exactly similar proposition to be true, which is begging the question. Many well-meaning men have published illogical arguments to prove the existence of a God, and it is fortunate that their failures have no logical effect upon the truth of that which they hoped to demonstrate.

185. I mentioned in the last Article that several philosophers had tried to prove a law of mechanics, but had begged the question by assuming some almost exactly similar proposition to be true without proving it. This **fallacy of begging the question consists in taking for granted that which has to be proved**, and is of great importance, because the fallacy is very difficult to detect and explain, and occurs in several different ways. Sometimes it arises from giving a name to a thing, and then supposing that we have explained the thing. A wise man, as

well as a child, may reasonably ask, why can we see through a glass window? Nobody has yet been able to give a reason why glass, crystal, and various solid things can be seen through, while most solid bodies cannot. But we sometimes hear it said that we can see through glass, "because it is transparent." This is clearly begging the question; to say a thing is transparent is neither more nor less than to say that you can see through it. The French dramatist Molière ridiculed fallacies of this kind very cleverly. The father of a dumb girl wants to know why his daughter is dumb. "Nothing is more easy than to explain it;" says the physician Ignarelle; "it comes from her having lost the power of speech." "Yes, yes," objects the father, "but the cause, if you please, why she has lost the power of speech." Ignarelle is quite ready with an answer. "All our best authors will tell you that it is the impeding of the action of the tongue."

186. The most frequent way, perhaps, in which we commit this kind of fallacy is to employ names which imply that we disapprove something, and then argue that because it is such and such, it must be condemned. When two sportsmen fall out in some matter relating to the subject of game, one will, in all probability, argue that the act of the other was unsportsmanlike, and therefore it should not have been done. Here is to all appearance a correct syllogism :—

No unsportsmanlike act should be done;
John Robinson's act was unsportsmanlike;
Therefore, John Robinson's act should not have been done.

This is quite correct in form; but it is evidently the mere semblance of an argument. "Unsportsmanlike" means what a sportsman should not do. The point to be argued was whether the act fell within the customary definition of what was unsportsmanlike.

187. People who do not like examinations are fond of saying that pupils are crammed for the purpose of passing them, and then they imply that the knowledge thus gained by "cram," is of little value. But this is very bad reasoning, and consists in falsely assuming that all or most candidates for examinations are crammed in the same way. If a pupil, being quite unable to understand a proposition in Euclid, learns it off by heart, and then writes it out in the examination room, as if he knew what he was writing, this is a bad case of cram, and the pupil gets no good beyond the exercise of memory. But if the pupil works up some books of Euclid, and can answer questions on them intelligently, he may have crammed them in the sense of doing it to pass the examination, but he has done it in a totally different way. Even though he forgets the problems in a few months or years, his mind will have been exercised in the best manner.

188. Words like "Cram" and "Unsportsmanlike," which are used in this fallacious way, have been called **question-begging epithets**, and we should always be on our guard against being misled by them. It is a good proverb which says "Give a dog a bad name and hang him."

XXVII.—FALLACIES IN INDUCTIVE REASONING.

189. I have already explained that the way in which people very commonly argue from one particular case to another is a very faulty and inaccurate mode of reasoning. It depends upon assuming that there exists some general resemblance or analogy between the cases, but in a great majority of instances people

make these inferences without taking the trouble to ascertain that there are sufficient grounds for what they do. People often disregard all precautions and assume that the medicine which suits one person will suit another, or that what cures one disease will cure another. **There is in all persons at all ages a tendency to hasty and false generalization.** The difficulty is not in making inferences, but in making correct ones. The mind is so framed that we cannot help classing together things which look like each other. The child does this as soon as it can speak a few words; it calls other men "papa" as well as its own father, because it has no clear idea of resemblances and differences between them. A beaten dog fears a stick even in the hands of a person who would never think of using it upon the dog. But persons with reasoning powers vastly greater than those of the child or dog often use them quite as faultily, and generalize in a very rash and careless manner.

190. Travellers sometimes make a rapid journey by railway through a foreign country, and then come home and write a book, as if they knew all about the country. They judge of millions of people by the few that they get to know slightly in hotels or public conveyances. If they are cheated by one or two people, they infer that most of the nation are dishonest. Too frequently we judge savage or partially civilized people from unfavourable specimens, with which alone travellers come in contact. The savages living on the shores of some unexplored lands, like New Guinea, have probably been ill-treated by the crews of trading vessels. Hence they are very unfriendly to strangers. But we ought not to generalize and infer, that all the inhabitants of a large country like New Guinea are exactly like those on the coast. Up to the present time, foreigners have not been able to travel safely in China, and can hardly visit more than Hong Kong,

Shanghai, Canton, Hang Kow, and a few other ports. We ought not to suppose that the whole of the vast population of China is like that with which we are acquainted in these towns.

191. There is really no good reasoning at all in assuming that other things or persons are like those which we have seen. In getting a sample of wine from a cask, as before explained (Art. 155), we know that it has been well mixed up, and, if requisite, we could mix it on purpose to make the sample a fair one. But we cannot mix up the population of a kingdom, and therefore we must not generalize about them unless we have seen so many persons, in different places and ranks of society, as to render it very probable that we have got samples of all the principal kinds. We should especially beware of judging about any people or place from newspaper reports of what happens. As people are most interested in reading about strange and serious events, such as murders, robberies, great accidents, riots, absurd deeds, and so forth, we frequently hear about these things, but not about the innumerable peaceful and every-day events of life. During the last few years, the newspapers of Manchester and Liverpool, have drawn attention to the savage way in which Lancashire men kick their wives and their friends, not to speak of unoffending strangers. Nevertheless, visitors from the more polished southern counties need not fear to meet a brutal kicker at every street corner. Fortunately, the kickers are still so small a proportion of the whole population, that we should hardly know of their existence but for the newspapers. Judging from the contents of American papers, especially as quoted in English papers, it might seem as if American gentlemen were constantly shooting their intimate acquaintances in bar-rooms; but, I suppose, a man may live in America all his life without seeing a revolver fired.

192. In this way trades-unions and societies of working men have been unfairly treated. Because some such societies have, at one time or another, employed people to commit illegal acts to punish workmen who broke the rules of their union, it is false generalization to speak of all unions as if they did the same. We cannot suppose that all working men, or all societies of working men, are exactly like each other, and it is most unfair to judge them all by the few worst cases which happen to be made public.

193. All the instances described in the last three articles are cases of false and hasty generalization. But we may without much difficulty distinguish three kinds of bad reasoning of the nature alluded to. Sometimes we argue wrongly that what is really true of a great many things, and, as a general rule, is also true of some special case which does not properly come under the rule. We extend the generalization too far. At other times we begin with that which is true only in certain special cases, and then treat it as if it were true of many things, and as a general rule. In the third place, we sometimes argue from one case which is peculiar, to another case which is also peculiar, so that there is no connection or real analogy whatever. These three kinds of fallacies we may then describe as (1) **from the general to the special**; (2) **from the special to the general**; and (3) **from the special to the special.**

194. It is a general law that all plants grow by absorbing carbon from the air under the influence of sunshine. If, therefore, we shut up a plant in a cellar, where no daylight can reach it, we should find it would not grow, as a general rule. But we must not apply this general rule to certain special cases, as, for instance, where a plant derives nourishment from a bulb, or tuber; potatoes, hyacinths, Jerusalem artichokes, and many like plants will sprout and partially grow in the

dark. Toadstools, mushrooms, and other kinds of fungi, again, are so different in many respects from flowering plants, that we should hesitate in applying to them any rule that has only been learned from the observation of flowering plants. A fungus is, in fact, capable of growing upon the carbon contained in the soil, and without the aid of light. Great quantities of mushrooms eaten in Paris are grown in caves under the town, and that delicate kind of edible fungus called the truffle grows altogether under the soil.

195. In legal matters we are frequently in danger of applying a law to cases which were not intended to come under it. Even when no special exceptions are mentioned in laws, bye-laws, or regulations, it may be evident that such exceptions exist. It is a very necessary regulation on railways that no one shall be allowed to jump out of a carriage in motion. But it is clearly understood that such a rule does not apply to railway guards, and other servants, who, by practice, can do it with much less risk than other people, and are often obliged to do it. Even a passenger would not be punished for breaking such a regulation, if he could show that there was more danger in remaining within the carriage than in jumping out, the only object of the rule being to save people from danger of injury.

196. Nothing is more clear in the laws of England than that no Englishman can become a slave, and a well-known song asserts in the most positive way that "Britons never shall be slaves." Yet the judges are continually occupied in sending persons into penal servitude, which is only a longer name for slavery. The fact, of course, is that the general rule about Britons is not intended to apply to the exceptional case of criminal Britons, though we seldom think of this when repeating the popular words.

197. The next kind of fallacy mentioned was that of **wrongly arguing from a special case to a**

general law. If we were to infer that, because arsenic, and strychnine, and prussic acid produce death when taken in considerable doses, they always produce death, we should be mistaken, because they are frequently given as medicines in exceedingly small quantities and much diluted. A large number of tee-totallers want to prohibit the sale of spirituous liquors altogether, and the reason which is sometimes given is that alcohol is a poison. It is quite true that when a large quantity of strong alcoholic spirit like rum or whisky, is drunk, it may produce death like a strong poison, and if taken frequently in too large quantities it is very injurious. But it is a fallacy to argue that it is therefore "poisonous" when taken in small quantities, and mixed with plenty of water. As I have already mentioned the most terrible poisons cease to be poisonous when taken in sufficiently small doses. It is all a question of degree and quantity.

198. There only remains to be considered that third kind of false generalization, which consists in arguing from one special case to another special case, between which cases there is no real connection. It would be absurd to argue that because a man when assaulted is justified in knocking his assailant down in self-defence, if he can do it, therefore one prize-fighter is justified in knocking down another. Each is a special case, and there is no true analogy between them. The practice of betting is sometimes defended on the ground that people are not blamed for speculating in cotton or corn. Why, then, should not people speculate upon horse races? The fact, however, is that speculation is not to be approved unless it brings advantages to the public. Speculations in corn, cotton, and other goods do, on the whole, bring advantages, both to the public and to those who make the speculations with the hope of profit. But speculation upon horse races does not bring any such advantages, and far more injury is done

to those who lose by betting than can be balanced by the profits of those who win.

199. It is not difficult to see that the fallacy here described as arguing from one special case to another is only a kind of fallacy of false analogy (Art. 167). But it is impossible too often to remind people that, on the one hand, **all correct reasoning consists in substituting like things for like things,** and inferring that what is true of one will be true of all which are similar to it in the points of resemblance concerned in the matter. **All incorrect reasoning,** on the other hand, **consists in putting one thing for another when there is not the requisite likeness.** It is the purpose of the rules of deductive and inductive logic to enable us to judge as far as possible when we are thus rightly or wrongly reasoning from some things to others.

QUESTIONS.

I.—INTRODUCTION. *Page* 7

1. How would you describe the purpose of the Science of Logic?
2. What does Logic teach us?
3. What did Lord Bacon say of knowledge?

II.—HOW WE COMMONLY REASON. *Page* 9.

1. How do mistakes in reasoning commonly arise?
2. Explain how we are likely to reason badly about putting a poker into the fire.
3. Can you explain, in like manner, how people sometimes reason wrongly when they say that iron is colder than wood?
4. What is a general law of nature?
5. By what kind of reasoning do we learn the laws of Nature?
6. What is deductive reasoning?
7. What is *to infer*?

III.—WHAT IS DEDUCTIVE REASONING? *Page* 12.

1. What is *to reason*?
2. In the following argument point out which are the terms, propositions, copulas, and signs of quantity:—

All Englishmen are subjects of the Queen.
All Kentishmen are Englishmen.
All Kentishmen are subjects of the Queen.

3. Of what parts is a proposition made up?

IV.—THE DIFFERENT KINDS OF TERMS OR NAMES. *Page* 14.

1. What kinds of words may a logical term consist of?
2. Explain the difference between singular and general terms.
3. What is a collective term?
4. How would you describe the nature of a concrete term and of an abstract term?
5. Look over the list of terms which you will find below, and point out, (1) those which are singular terms; (2) those which are general; and so on in succession those which are (3) collective; (4) concrete; (5) abstract; (6) positive; (7) negative.

House.	Friend.
London.	Friendly.
School.	Friendliness.
Scholar.	Roger Bacon.
Impossibility.	Parliament.

6. What are the negative terms which belong to the following positive terms?—

Full.	Absent.
Faithful.	Limited.
Free.	Intentional.

7. What is the negative term of *dwarf*? What is the opposite term of dwarf? Explain the difference between the negative and the opposite.
8. What are the opposite terms belonging to (1) gain; (2) right; (3) up-hill; (4) advancing?

L. P.

V.—THE FULL MEANING OF TERMS. *Page* 20.

1. What do you understand by the extension of the term *printing-press?*
2. How does *printing-press* differ in extension from *steam printing-press?*
3. How do the terms *rose, moss-rose,* and *flower*, differ in extension?
4. What do you mean by the intension of a term?
5. Is the intension greater in the case of *printing-press* or of *steam printing-press?*

VI.—THE CORRECT USE OF WORDS. *Page* 22.

1. What is the meaning of a word?
2. When is the meaning of a word ambiguous?
3. After a little consideration mention some of the various meanings with which we use the word *box*. Do the same with any of the following words:—

Square.	Seal.
Cloth.	Letter.
Day.	Morning.

4. Say what you mean exactly when you speak of a *house*.

VII.—HOW AND WHY WE CLASSIFY THINGS. *Page* 27.

1. What do we do when we classify things?
2. What is the use of classifying things?
3. What useful facts do we know about the following classes of things? (1) grasses; (2) non-conductors of heat; (3) arteries of the body; (4) oils.
4. How does a *genus* differ from a *species?*
5. In the case of *pen* and *steel-pen*, which is the genus and which the species?
6. Point out which of the following classes of things may be called a *genus*, when compared with some one of the others as a *species*:—

Books.	Spar.
Red ink.	Dictionaries.
Vegetable.	Liquids.
Ink.	Bibles.

7. Try if you can make a perfect division of the class *picture*.
8. Has a species a greater or less intension than the genus?
9. What is a logical difference?
10. What do we mean by a definition?
11. Try if you can define the meanings of the terms *crystal, triangle, fibre.*
12. What is the logical difference between *pen* and *steel-pen*, and similarly between *man* and *giant*, *cathedral* and *church*, *money* and *coin?* You must first make out which is the genus and which the species.
13. Mention some logical accidents of the following classes: book, carriage, tree, pianoforte.
14. Mention some logical properties of the following classes: horse, swallow (bird), grass, triangle, organ.

VIII.—PROPOSITIONS. *Page* 37.

1. Examine the following propositions, and point out in each where we are to look for the subject, copula, and predicate:—

St. Paul's is the Cathedral of London.
Manners makyth man.
Time flies.
"True, 'tis a pity: pity 'tis, 'tis true."

2. How does a negative proposition differ from an affirmative one?
3. What is a hypothetical proposition? Give some examples of such propositions.
4. How do you recognise a disjunctive proposition?
5. Pick out of the following propositions, (1) those which are *universal*,

and then in succession, those which are, (2) particular, (3) negative, (4) hypothetical, (5) disjunctive:—

If the barometer rises, it will not rain.
All rivers flow downwards.
No milk-maids have small-pox.
Most who die of small-pox are not vaccinated.
Some electors are unfit for their duties.
The most convenient books are octavos or duodecimos.

6. What is the quantity of the predicate in the propositions?—

Some well-water is poisonous.
Some costly ornaments are not beautiful.

7 Draw circles showing how the subject of each of the following propositions is included or not included in the predicate:—

Some roses are yellow.
All male ratepayers are electors.
Some ratepayers are not fit to be electors.

8. What are the quantities of the subject and predicate in a universal negative proposition?

9. Make an example of a universal negative proposition, and show how it may be represented by two circles.

10. Explain exactly what you suppose me to mean when I say:—

Some honest acts are wise acts.
Some games are dangerous.
Few murderers escape.

11. Show the exact meaning of each of the above propositions by drawing circles with broken lines.

IX.—HOW TO CHANGE PROPOSITIONS. *Page* 47.

1. How can we convert the propositions:—

Some roses are yellow.
No men are perfectly wise beings?

2. By what kind of conversion can we convert universal affirmative propositions?

3. Convert the propositions:—

All bricks are made of clay.
All steel is substance containing carbon.

4. Represent by circles the propositions:—

All dictionaries are books.
Not all books are dictionaries.
Some books are dictionaries.
Some dictionaries are books.
Honesty is the best policy.

5. "All soot is black." What can I infer from this proposition about things which are not black? What do I learn about things which are not soot?

6. If we know that all acts which are not honest are not wise, what proposition can we thence infer with wise acts as the subject and honest acts as the predicate?

X.—SYLLOGISM. *Page* 53.

1. All animals want food.
Oysters are animals.
Therefore, oysters want food.

After examining the above syllogism, answer these questions:—

(1) Which is the conclusion?
(2) What are the two other propositions called?
(3) Which is the middle term?
(4) How do you know it to be the middle term?
(5) Which is the major term? Why?
(6) Which is the minor term? Why?
(7) What is the name of the proposition containing the major term?
(8) How would you represent the syllogism by circles?

2. Answer the same questions as regards this syllogism :—

Many houses are unfit to live in, because they are badly drained, and all badly drained houses are unfit to live in.

XI.—THE RULES OF THE SYLLOGISM. *Page* 56.

1. How many terms are there in a syllogism? Why cannot there be more?
2. How many propositions must there be in a syllogism?
3. What is meant by the middle term of a syllogism being *distributed?*
4. Explain how it is that the middle term is not distributed in the following false argument :—

All Red Indians are uncivilised people; for all Red Indians live by hunting, and some uncivilised people live by hunting.

5. Why can we make no inference when both the premises are negative?
6. What kinds of premises must we have in order that we may get a negative conclusion?
7. Write out from memory all the six rules of the syllogism.
8. What is the minor rule about two particular premises?
9. If one premise be particular, what do you know about the conclusion?
10. Draw circles to represent this syllogism :—

All birds lay eggs.
No bats lay eggs.
No bats are birds.

11. What conclusions can I draw from the following pairs of propositions?—

{Gold is a metal.
{No metals are compound substances.

{All glaciers consist of ice.
{All glaciers move downwards.

12. After making syllogisms with the above pairs of propositions, draw circles to represent the syllogisms.
13. Explain how it is that the fourth rule of the syllogism is broken in this false argument :—

All sea water contains salt; but the water of Lake Superior does not contain salt, for it is not sea water.

14. What rules of the syllogism are broken by the following false arguments :—

This is an animal: for it can move, and all animals can move.

Some things attained without labour are valuable; some knowledge is attained without labour, and is therefore valuable.

Marcus Aurelius was a good man; Marcus Aurelius was a Roman Emperor; therefore all Roman Emperors were good men.

XII.—HYPOTHETICAL SYLLOGISMS. *Page* 69.

1. What are the rules of the hypothetical syllogism?
2. Show how the following arguments break those rules :—

(1) If John Jones has a summons he appears at the county court; but John Jones does appear at the county court, therefore he has a summons.

(2) If the Americans have much gold coin they are prosperous; but they have not much gold coin; therefore, they are not prosperous.

3. Show that the first of the above arguments (1) breaks the third rule of the ordinary syllogism, and the second (2) the fourth rule.
4. If water be warmed it will evaporate. This water is warmed; therefore, it will evaporate.

Make the above argument into an ordinary syllogism.

XIII.—OTHER KINDS OF ARGUMENTS. *Page* 71.

1. How do you know a disjunctive proposition?
2. Make some examples of disjunctive propositions with two or with three alternatives.

3. What are the rules which we must obey in making a disjunctive syllogism?

4. Which of the following arguments obey the rules of the disjunctive syllogism?—

Melodies are either in a major or a minor key; this melody is not in a major; therefore, it is in a minor key.

Beautiful flowers are liked either for their scent or their appearance. Roses are beautiful flowers, and are liked for their scent; therefore they are not liked for their appearance.

XIV.—THE GREAT RULE OF INFERENCE. *Page* 73.

1. What is the Great Rule of Inference?

2. How do I make an inference when I argue that because a seal is a warm-blooded animal, therefore the heart of a seal is the heart of a warm-blooded animal?

3. If honesty is the best policy, it follows that what is not honesty is not the best policy. Show how the above agrees with the rule of inference.

4. All coal contains carbon: why may I not infer by the rule of inference that what is not coal does not contain carbon? What may I infer about things which do not contain carbon?

5. What do you understand by the Substitution of Similars?

6. How do we make a substitution when we argue that because John is of the same height as James, but William is taller than John, therefore William is taller than James?

7. If John is of the same height as James, and Walter of the same height as John, *how* can I infer by substitution that Walter is of the same height as James?

8. If William is taller than John, and Robert is not of the same height as William, what can I infer about the heights of Robert and John?

XV.—INDUCTIVE REASONING. *Page* 76.

1. How would you describe deductive reasoning?
2. What is the difference between inductive and deductive reasoning?
3. What was Galileo's method of acquiring knowledge?
4. What was Galileo's celebrated experiment?
5. At what times respectively did Roger Bacon, Galileo, Lord Bacon, and Sir Isaac Newton live?
6. What is the Novum Organum?
7. What is the first step in inductive reasoning?
8. What is the second step?
9. What is the third step?
10. What is the fourth step?
11. Describe what is meant by a hypothesis.
12. When we have made a hypothesis, what are we to do with it?
13. What were the several hypotheses made to explain the origin of fossils? How were they disproved or proved?
14. In what way did Newton argue about the Law of Gravity?
15. How was Descartes' hypothesis disproved?
16. What is a Crucial Instance?
17. What is an Experimentum Crucis?

XVI.—INDUCTIVE REASONING IN ORDINARY LIFE. *Page* 85.

1. If I find in the morning a branch broken from a tree in my garden, and it might be caused either by a high wind in the night or by a thief, how may I proceed to ascertain which is the case?

2. Why are apparently unimportant facts sometimes the best evidence?

3. In what way was it made very certain that the Claimant was not Sir Roger Tichborne?

XVII.—OBSERVATION AND EXPERIMENT. *Page* 89.

1. What is the difference between experimenting and merely observing?
2. What are the advantages attaching to experiment?
3. How are we most likely to discover new substances?

XVIII.—ANTECEDENTS AND CAUSES OF EVENTS. *Page* 92.

1. What do we mean by the antecedents of an event?
2. What are consequents?
3. How do you distinguish *causes* from other antecedents?
4. Supposing a house to be burnt down, mention all the antecedents and consequents you can think of, and point out any of the consequents which might be causes.
5. What is the great rule of experiment?
6. The carpet of a room becoming faded, how would you attempt to ascertain whether this is due to the air, or the sun, or the fire, or people walking on it, or by some cause of change in the carpet itself?

XIX.—DISCOVERY OF AGREEMENT. *Page* 95.

1. What is the first step towards discovering the causes of any kind of event or appearance?
2. Supposing a person to be ignorant why beer froths up when poured out, with what other things should he compare it to try and find out the cause?
3. The sun when setting is sometimes of a deep red or orange colour. In what other cases do we see a similar colour and appearance?

XX.—THINGS WHICH VARY IN QUANTITY. *Page* 97.

1. What is the rule for experimenting upon things which vary in quantity?
2. How are we to recognise the cause or effect of a thing which varies in quantity?
3. A healthy person who walks briskly in a sheltered place does not feel cold even when the air is very cold. How would you ascertain by experiment whether his feeling warm is due to exercise, or to shelter, or to both?

XXI.—THINGS WHICH VARY PERIODICALLY. *Page* 99.

1. What is periodic variation?
2. Mention some instances of periodic variation.
3. How can we recognise the effects of a periodic cause, or the cause of a periodic effect?
4. If on looking into a manufactory you see two wheels, one of which always moves or stops when the other moves or stops, what would you infer about them, and by what mode of reasoning?

XXII.—REASONING FROM EXPERIMENTS. *Page* 102.

1. What is the general rule in reasoning from one experiment to another?
2. Striking a flint and steel together, I produce a spark. How might I proceed to ascertain by experiments and reasoning what other substances would be likely to produce the same results?
3. What is to generalize?

XXIII.—HOW AND WHEN TO GENERALIZE. *Page* 104.

1. What do you mean when you speak of a sample or specimen of a class of things?
2. Under what circumstances can we infer from a sample without having the things themselves?

3. What is the rule of generalization about things which resemble each other in many properties?

4. What is requisite in generalizing about things which resemble each other in few properties?

5. Seeing a wire stretched across between two houses, can you tell how to infer whether it probably is or is not a telegraph-wire?

6. Why is reasoning by analogy so uncertain?

7. There is more printing in a copy of the *Daily News* than in this Logic Primer. Might we not then infer by analogy that it would be profitable to sell Primers at a penny each?

8. If letters can be carried for a penny each, why cannot the Post Office profitably convey telegrams for a penny each?

XXIV.—REASONING BY ANALOGY. *Page* 107.

1. What is reasoning by analogy?
2. How does it differ from reasoning by generalization?
3. What is the rule of reasoning by analogy?
4. Say as exactly as possible how you would ascertain whether a piece of metal which seemed to be gold was really so, or not.
5. When Robinson Crusoe saw a foot-print in the sand, why and how did he infer that there was another man on the island? Answer this question carefully.

XXV.—FALLACIES. *Page* 112.

1. What are the two meanings of the word fallacy?
2. With which meaning ought we to use the word in Logic?
3. Mention all the fallacies which arise from breaking the rules of the syllogism.

XXVI.—FALLACIES OF AMBIGUITY. *Page* 114.

1. Mention some cases in which the *right* is *wrong*.

2. Explain how the fallacy of ambiguity arises in each of the following arguments:—

(1) Railway travellers are subject to frequent accidents; I am a railway traveller; therefore I am subject to frequent accidents.

(2) Men who do no work are idle; this man is not doing work; therefore, he is idle.

(3) Men are numerous; kings are men; therefore, kings are numerous.

(4) A man who drinks is very culpable; this man is drinking; therefore, he is very culpable.

(5) Riding is pleasant; he is riding; therefore, he is pleasant.

3. How does the failure of an argument properly affect the truth of the conclusion?

4. What are *question-begging epithets?* Can you think of some examples?

XXVII.—FALLACIES IN INDUCTIVE REASONING. *Page* 122.

1. What is the kind of reasoning which we notice in the lower animals? What is the defect in it?
2. How are we likely to be misled by travellers' tales and by newspaper reports?
3. What are the three kinds or cases of fallacious reasoning by generalization?
4. The Nile by overflowing its banks enriches the neighbouring country; therefore, the Po by overflowing its banks will enrich the neighbouring country. What do you think of this argument?
5. If correct reasoning is always the substitution of similars for similars, how would you describe incorrect reasoning?

Richard Clay and Sons, Limited,
LONDON AND BUNGAY.

Made in the USA
Monee, IL
04 November 2026

67385012R00079